EAST EUROPEAN CARS

JULIAN NOWILL

SUTTON PUBLISHING LIMITED

Sutton Publishing Limited
Phoenix Mill · Thrupp · Stroud
Gloucestershire · GL5 2BU

First published 2000

Title page photograph: The GAZ A Aero, built in 1934 and designed by A.I. Nickitin.

British Library Cataloguing in Publication Data
A catalogue record for this book is available from the British Library.

ISBN 0-7509-2369-5

Typeset in 10.5/13.5 Photina.
Typesetting and origination by
Sutton Publishing Limited.
Printed and bound in England by
J.H. Haynes & Co. Ltd, Sparkford.

The very rare military version of the Wartburg 311 is shown here in a suitable period setting.

Contents

UNLOVED SOVIET SOCIALIST REGISTER

СПИСОК НЕЛЮБИМЫХ СОВЕТСКИХ И СОЦИАЛИСТИЧЕСКИХ МАШИН

The USSR club was formed in 1991 and caters for enthusiasts of Eastern Bloc cars, trucks and bikes. Contacts supply hard to obtain spares and there is an annual tour of Russia including the Moscow Motor Show. For information, contact the USSR, Earlsland House, Felix Djerzinski Straße, Bradninch, Devon, EX5 4QP, or email nowillj@greigm.co.uk.

The design of the GAZ 18, which was created in 1958 by engineers in Gorky, is entirely original. Of the two cars that were built, only one exists today. It has a 2-cylinder 500cc engine with a hydraulic clutch and hydraulic transformer – a prototype automatic gearbox. This allowed the clutch pedal and the usual gearlever to be dispensed with: in fact, the vehicle was intended for use by invalids. The petrol tank and boot are at the front. This GAZ 18 was given to the Gorky Automobile Factory Museum by the magazine *Autoworld* on 4 August 1999 during the fourth Moscow Motor Show.

Introduction

Why do people become car bores? Some want to make new friends, others are so inadequate they do not have any friends to start with, while the bird-pulling aspect is not to be ignored. Status? Investment potential? Owning an Eastern Bloc car will give you none of these things. While values are rock bottom, so is investment potential, and when one has grappled with an owner's manual in Cyrillic and waited six months for some brake master-cylinder rubbers, one will appreciate that commie car ownership is not for the faint hearted.

Many fallacies are perpetuated about Eastern European cars such as the fact that Trabants are made out of cardboard (they are not, rather a form of formica which is strong and durable). Eastern European cars lack the power to weight ratio of machinery in the West, but then until recently there were no motorways in Eastern Europe. However, Soviet Bloc cars come into their own when bouncing along ruts north of the Arctic Circle while trying to digest 76 octane fuel. For the DIY motorist, they are a pure pleasure and with TLC will last far longer than Western kit of twice the price. Sadly, British buyers tended to be those on the breadline and, as a result, their steeds often received no maintenance whatsoever – which is why low mileage cars with perhaps one minor fault can be picked up from scrap dumps for less than the price of a pair of jeans.

Car designers in the East had to put up with serious disadvantages such as lack of investment, raw material shortages and, in some countries, diktats from the communist authorities forbidding the manufacture of motor vehicles. As a result, many small-scale operations in Czechoslovakia and Hungary produced 'invalid' or three-wheeled motorcycles.

Digging deeper, you will find that Eastern Bloc machinery is brilliantly designed. How about the Moskvich 427 where everything is adjustable and there is even a glass viewing panel on the carburettor to check the fuel level. Most Ladas have hydraulic headlamp adjusters so that the beam can be adjusted for load from the driver's seat. Even the lowly Trabant has a sophisticated intermittent wiper setting which can cope with anything from mist to a light downpour.

It is unlikely that Eastern Bloc machinery will ever be loved or cherished by the typical classic car buff but perhaps this is a positive advantage, with prices starting from £100 for a rear-engined Skoda to £400 for an MOTd Trabant. The top end of the market does, obviously, exist with the best Tatras fetching over £25,000. What is the ultimate in Eastern Bloc wheels? This must be the MAZ 543 Scud missile launcher with twin petrol V8 engines and central tyre inflation.

Acknowledgements

Writing an acknowledgements section is a dangerous business as there are many people whom through my general disorganisation I will fail to credit for their help. As I cannot type, a big vote of thanks to the girls of Greig Middleton who worked in their lunch break: thank you Jill Pritchard, Louise Denning, Lisa Holman and Sandra Martin. Holger Windolph translated large quantities of German text to English, and also drove the Syrena featured in the book from Poland to England.

When researching the book, my thanks to (in no particular order) the RAC motor library, Pall Mall, London and Beaulieu motor library. Thanks to Margus Kuuse from Estonia, for his help. Although I failed to ever make contact, Glen Lundqvist from Sweden has published unique research, which I have found very helpful. For the East German chapter, my thanks to Wolfgang Melenk (Hamburg). For Hungary, thanks to Pal Negyesi, the founder of Budapest's *Autoclassic* magazine. For the Ukraine, my thanks to Dennis Lesley for information on the Tavria and for the Polish chapter my thanks to Robert Mackay. For the Czech Republic and Slovakia, my thanks to Marian Suman-Hreblay, and Greg Merksamer for pictures. The photograph for the cover was kindly produced by Karin Haberle from Leipzig, founder of Germany's Chaika club. Many photographs which have appeared were simply posted to me and, as I failed to keep a record of the senders, my thanks to those unnamed photographers.

A very large vote of thanks indeed to Ed Hughes, who effectively wrote the Wartburg, Moskvich and Lada chapters – editor of the Wartburg/Trabant owners' club magazine, and well known kerbside mechanic in East Devon. He has come on two trips to Russia with me, and his enthusiasm and knowledge (particularly on Russian door handle design) verges on the obsessive.

Finally, my thanks to Simon Fletcher at Sutton Publishing for the idea of this book. Despite my insisting that it would a non-runner, he persuaded me otherwise and the fact that you, dear reader, have bought this humble offering means, hopefully, that Sutton will at least recoup the cost of their telephone calls and helpful letters to me.

1

Russia

The first mention of GAZ was in Russia's first five-year plan of 1928 to 1933 when Gorky was still known under its original name of Nizhny Novgorod. On 1 January 1932, two years after the foundation stone was laid, series production of NAZ AA started; it was very closely based on the Ford Model A. Almost immediately NAZ changed its name to GAZ when the town was renamed after the Russian poet Gorky.

GAZ also enjoyed success with the GAZ A lorry, which was produced in far greater numbers than the car. Because of the poor roads, the typical Russian transport was often a lorry rather than a car because of its toughness. It made little difference to the driver whether he was transporting peasants or potatoes. One of the first Russian long distance rallies was the Moscow–Karakum 6,200 mile rally where the GAZ lorry was the most numerous participant, although six GAZ A cars also took part. In 1936 GAZ launched the M1, seen here, a 3.3-litre machine produced until 1942. GAZ engineers cooperated with AMO/ZIL in Moscow on development and there were even a limited number exported to France in 1937. The successful M1 came in a number of variants, including a half-track, and production totalled 62,888. The GAZ M1 was developed into the GAZ 11 in 1940 with a 6-cylinder engine. Because of the war only 1,250 were made, with GAZ generally producing military trucks such as the GAZ 64, 67 and 61. After the war GAZ 11 production was increased. The model finally died in 1948.

During the war the main GAZ engineer, Lipgart, pleaded for funds to develop a new model and on 9 May 1945, one day after VE Day, the GAZ M20 Pobieda was launched, Pobieda meaning victory in Russian. The common view in the West is that Russian engineering is always 'old hat' but the Pobieda must have caused quite a stir at the end of the war being one of the world's first monocoque cars. Looks were reminiscent of the Standard Vanguard phase 1 and the side-valve engine was a good (if thirsty) slugger.

From 1946 to 1958 a total of 235,997 Pobiedas were built, including 14,220 cabriolets and the four-wheel drive M72 produced between 1955 and 1958. Engineers at GAZ played with a number of alternatives including the GAZ 51, a 6-cylinder Pobieda which never made it to production. The automotive engineering institute NAMI in Moscow developed a booted version which remained in prototype form (the Poles produced a booted version many years later). Most exciting of all was the Torpedo (above), a 1951 streamlined machine capable of 118mph, although Soviet Arctic explorers would have preferred the aero-engined Poby sledge (below). Developments to the Pobieda were restricted to a change of grille, and there were limited exports to Scandinavia.

Today the Pobieda is gradually becoming a collector's car in Russia, although some are still in daily use in poorer areas. Many have had the ohv Volga engine and gearbox retro-fitted to improve fuel economy. Driving the Poby is a surprisingly relaxed affair. You sit high as in a Range Rover and while the view through the rear mirror is not marvellous, because of the fastback design, it is adequate for parking. Behind the front bench seat rear legroom is extensive, and the brakes are good, no doubt helped by the retro-fitting of a GAZ servo.

A replacement for the Pobieda was planned as early as November 1953 when the chief engineer at GAZ, Alexander Hevsorov, sketched a design which combined ideas from a number of Western manufacturers such as the American Ford Customline and even perhaps the British Ford Zodiac. Early Volgas had ten vertical slits in the bonnet and a chromed deer mascot, which represented part of the coat of arms of Nizhny Novgorod where the GAZ factory was situated. The engine was all-new, an ohv design mated to a three-speed manual or three-speed automatic box. Gearboxes until 1957 were very bulky, projecting into the rear passenger space. Independent front suspension used the old lever-type shock absorbers and rear suspension was longitudinal half elliptic springs. Tyres were 15in. A radio with two ivory knobs was standard, although taxi drivers were miffed that these were deleted from taxi versions. To add further to their chagrin, the Volga taxi did not have a cigarette lighter – although it did have an ashtray. Factory testing began on 3 May 1955 with one manual and two automatic cars. During the test, which was followed by the magazine *Ogonyok*, the three machines were driven to the Crimea and back from Moscow. Particularly impressive was the way the big leaf springs swallowed up ruts. A centralised lubrication system appeared on early cars, with oil passing through brass tubes to nineteen lubrication points. An occasional press on the pedal activated the system, but the oil deposits left under parked cars were not popular with pedestrians. As the pipework was often damaged by stones, this option was soon deleted. Because of problems with the ohv engine, the first three production Volgas had the old Pobieda side-valve engine (65bhp), and these rolled off the production line in October 1956. In 1957 full production began, although the Pobieda was still built for a further year. A five-pointed Soviet star crowned the new grille.

Finally, in October 1957, the ohv engine was introduced. The engine was an all alloy unit and the capacity of 2445cc continues to this day. Bhp rose to around 70 at 4000rpm. In 1958 automatic transmission was abandoned, following problems with the oil quality and lack of maintenance expertise at Russian garages. The car was shown to the world at World Expo in Brussels in 1958, where it was awarded a Grand Prix. The Belgians nicknamed it the tank on wheels while the Finns called it a tank in evening-wear. A right-hand drive version (GAZ 21P) was launched, with cars being sent to Indonesia and Britain; one even made a small appearance in the James Bond film *Thunderball*. The Volga on the left was driven to the UK by two Ukrainian chancers. After trying to sell it to the author for £3,000, they settled for an MOT-failed Lada and some dollars, and drove home happy.

The GAZ 12 is more commonly known as the ZIM (Molotov Factory). Although the GAZ factory should have been known as the Molotov works from 1932 to 1956, the name never really took on except for this seven seater, which GAZ sold to officials who did not measure up to a ZIL. The 3½-litre 6-cylinder engine came straight out of the GAZ 51 lorry and with twin carbs produced 89bhp at 3600rpm. With a clear American influence, the ZIM was 18ft long, with independent front suspension and a live rear axle. Top speed was 75mph, and the only variation was a landaulette. The ZIM was never popular, but understandably so as it was an up-market vehicle. Many ZIMs were used as people carriers, given the lack of a Soviet minibus until the RAF arrived. ZIMs were never officially exported, although one clocked up the most colossal mileage as an Istanbul taxi before being retired in the early 1990s. ZIM production lasted from 1950 to 1959, and the styling is well proportioned – from the majestic egg-crate grille to the chrome spear at the side and the self-effacing rear with modest stoplamps. In 1950 it was bang up to date with the best that Detroit could produce; however, by 1959 the Americans had moved on. For example, split screens had long since been replaced by curved glass, the technology for which was still lacking in the USSR.

While the ZIM was self-effacing, the Chaika (which translates as Seagull) was most certainly not – and this flashy amalgam of Detroit and Gorky is not for the shy. Towards the end of the '50s GAZ engineers looked at a restyle for the ZIM, and decided to start with a completely blank sheet of paper. Continuing ZIL's love affair with Packard, they produced a giant machine with more than a passing resemblance to the 1955 Packard Patrician. An all-aluminium alloy V8 engine of 5.526 litres drove through a push-button three-speed automatic transmission. As with the ZIL, windows were electrically operated and there was seating for seven with two occasional seats in the rear. Although Chaikas were produced until 1976, production probably never exceeded 2,000, and a surprisingly large number are finding their way westwards.

In 1976 the frivolous styling of the GAZ 13 was thought to be out of place and the designers produced a ZIL-class car without the glass partition. The seven-seater body is 8in longer than the Chaika and 15in longer than the Daimler Limousine. The engine needs its 220bhp to carry this 2½-ton car from rest to 62mph in 15 seconds. Top speed is, supposedly, 109mph. Between 1977 and 1988, 1,114 were produced – until the changed political climate signalled the end of limo production.

The GAZ 24 Volga is reminiscent of the big barge-like Chevrolet Impalas of the early '60s or, in Britain, the boxy Ford Zephyrs and Zodiacs beloved of the 1970 executive with his plastic briefcase. Compared to the GAZ 21 Volga, the wheelbase expanded from 8ft 9in to 9ft 1in while the height shrank from 5ft 3in to 4ft 10in. Road clearance also fell from 7½in to 7in, although only 110lb of weight was shed. The same engine was used, but maximum speed rose from 80mph to 90mph, thanks to a higher compression ratio (8.2:1 as opposed to 6.1:1). The three-speed transmission was replaced by a four-speed, and with the change of tyre size from 15in to 14in a new axle ratio was employed. Brakes finally had servo assistance and the thirty-seven lubrication points were reduced to nine. One similarity to the British Bristol was the fact that rear demisting was achieved by fan rather than heating elements. A double-barrelled carburettor replaced the old version and floor shift replaced column shift. In 1970 the GAZ 24 was a reasonably up-to-date proposition in the West, and Volgas were shipped to Belgium for a diesel engine to be inserted for taxi drivers.

Developments to the basic GAZ 24 Volga started with the GAZ 2410, which had an improved carburettor to help fuel consumption. In 1982 the GAZ 3102 was launched, with bigger rear lights and a host of other improvements. In 1986 the Volga 3807 was launched; it was a LNG taxi variant of the 3102. There was even a diesel model 4301 with a 74bhp diesel engine, while NAMI developed a turbo version. Nothing came of this, however. The year 1986 saw a modest restyle with the relaunch of the 3102 model, seen here. Buyers had a choice of engine power – 90bhp from the 8.2:1 compression ratio, 97bhp from the 8.1:1 three valve per cylinder engine, or 104bhp from the highest compression ratio. Assembly in Belgium continued, and the two diesel engines (2.3 and 2.0 litres) were destined purely for Europe.

The main model in current production is the GAZ 3110. Production started in 1996, after over half a million of the previous model had been built.

Disc brakes appeared at last, although only on the front, and a five-speed transmission is available as an option – as is a 16-valve 150bhp engine. The colossal boot was shrunk by 4in with the boot lid descending to bumper level. In the interior a rev counter was now standard, and seats were provided with variable lumbar support. Estate cars are not popular in Russia, for reasons of fashion – and many are used as ambulances. Note the airdam at the back, which is intended to keep the rear window free of mud. In true Russian tradition, the original 3102 model remains in production.

Throughout the '90s Russia was without a locally produced limo, and although ZILs were still being manufactured the 'campaign against privilege' meant that a less ostentatious machine was called for. The 3105 is fairly anonymous, with faintly Audi-like looks. The engine is a 3.4-litre V8 with a 0–60mph time of 11 seconds and a maximum speed of 125mph. On the inside there are electric seats (with memory settings for the driver), air conditioning, ABS and power steering, while the centre console contains mobile phone, ashtray and switches for the heated seats. Petrol consumption is 19mpg, which is not bad given that four-wheel drive is standard. It would appear that the 3105 has died a death, possibly because the price of US$90,000 was too high.

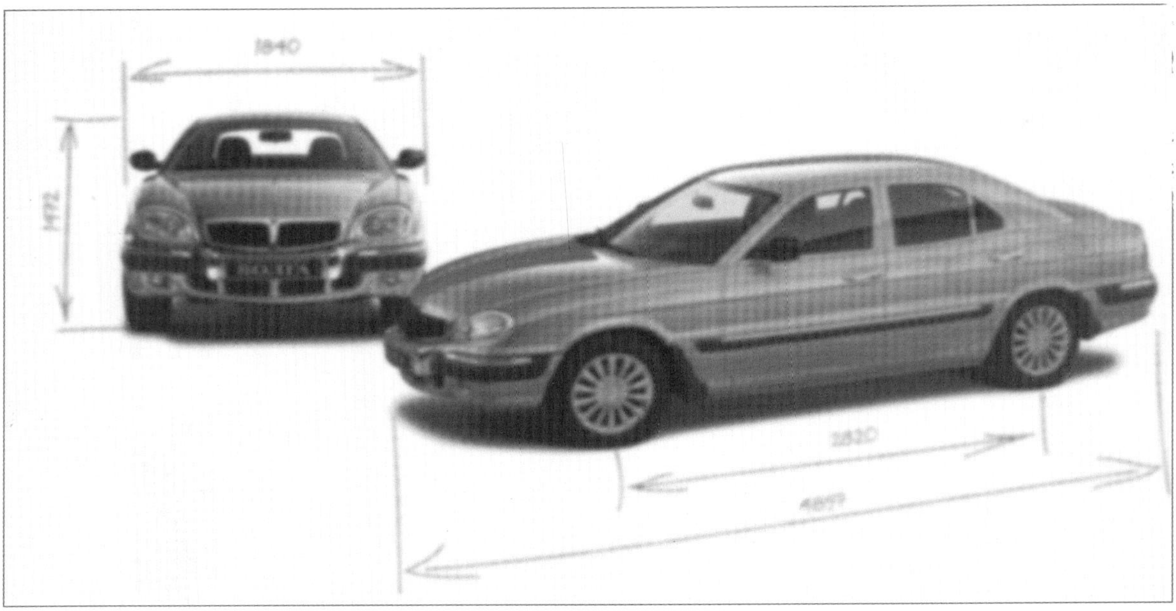

Thanks to the colossal success of the Gazelle, GAZ's Ford Transit look-alike pickup, GAZ has been able to finance a number of interesting projects. Ready for production is the new Volga 3111, on display at the 1998 and 1999 Moscow Motor Show. While this is a competent well-screwed-together machine, the typical Russian dislikes paying more than £3,000 for a new car and thus it may be many years before production is reached – by which time the looks will have started to fade.

Perhaps more promising is the launch of Russia's answer to the Dodge Ram, namely the GAZ 2308 Ataman, with production starting in 2000. GAZ does not have a sports utility vehicle in its line up, and one thought is to turn the Ataman into a passenger vehicle.

GAZ has already commissioned an American engineering company, Venture, to produce a stylish SUV on the Ataman chassis. Power will either be the good old 4-cylinder petrol or 5-cylinder diesel, manufactured under licence from Steyr. In the first seven months of 1999 GAZ produced 57,000 trucks and 73,000 cars (the limit of production). The future of GAZ as an independent car producer must be good. The fact that Russians refuse to pay top dollar for their cars means that foreign manufacturers simply cannot make a profit, and the production of big brawny workhorses ideal for local conditions may continue for many years to come.

During the Second World War the Russian Army urgently needed a Jeep-type vehicle, and within a very short space of time the GAZ factory produced the GAZ 67. This used the Model A engine, and had very basic styling with no compound curves anywhere. Production in wartime was limited, and these early vehicles can be identified by a grille made of welded poles (post-war vehicles have a proper stamped grille). Production totalled 4,851 between 1943 and 1945, and ceased in 1953 with chassis number 92,843. The model had its role in the Korean War, and was nicknamed HBW by the Red Army (translated as 'Want to be Willys').

The GAZ 67 above was photographed at the Beltring military vehicle festival in Kent, while the GAZ 69 on the left was caught in Russia in front of a UAZ four-wheel drive. Note the cut-out from the rear wheel designed to accommodate a spare wheel – which also explains the curious shape of the door. Also note the military spec. headlamps. The GAZ 69 below featured in the film *Empire of the Sun* and has non-standard headlamps. After being blown up for the film, it languished in a Devon scrapyard before being restored by its current owner.

Development of the four-wheel drive GAZ 69 took place in 1952: the design was all new with the engine being taken from the GAZ M20 Pobieda. The GAZ 69 was a great success with the Red Army and also saw civilian use; 634,256 machines were produced before 1972. Although production shifted in 1954 to UAZ, the UAZ 69 name was rarely adopted. Production in Gorky continued until 1956 and a number of variants were made. Particularly noteworthy was the GAZ 46 amphibian, which could manage 6mph on water from a small propeller at the rear, steering being by the front wheels.

Like most other Russian car manufacturers, UAZ is within a stone's throw of the River Volga, the plant being based in the attractive city of Ulianovsk. The plant opened in 1941, initially producing heavy trucks. In 1954, however, GAZ shifted production of its four-wheel drive GAZ 69 to the plant. Much to the irritation of the good burghers of Ulianovsk, the great Russian public continued to refer to this favourite four-wheel drive as a GAZ.

In the early '70s the UAZ engineers decided on a re-skin for the old UAZ 69, the 469, and came up with a more modern body, retaining the Volga-derived engine, leaf springs and general design – including the rather curious design of the front door, the unnecessary cutaway enabling front and rear doors to be reversed. The steel body is mounted on a conventional ladder-frame chassis with a wheelbase of 9ft 2in. Semi-elliptic leaf springs with beam axles soften the blows at the front and rear, while damping is by lever arm hydraulic shock absorbers all round. The de-tuned Volga engine produces 70bhp at 4000rpm, driving through a four-speed primary gearbox with a two-speed part-time transfer case. The interior is as spartan as only a Soviet military vehicle could be. Bare metal predominates and seats are trimmed in a tough vinyl material. There are two front seats with a three-piece foldable bench behind, and finally two further inward facing individual seats in the rear. Under axle ground clearance is 8½in, and approach and departure angles are a very respectable 50 and 45 degrees respectively.

Today the UAZ 469 remains in production, although confusingly the range has been renumbered. The base model is identical to that which rolled off the production line on 15 December 1972, and is known as the 31512. Coil springs are now an option and 92bhp is in line with current Volga engines. An estate car version is also available, the 31514, while top of the line is the 31513 long wheelbase with 100bhp engine, coil springing and power steering. Top of the range is perhaps a misnomer, although on the inside at least the naked steel is now covered in plastic. There is even carpeting on the floor, and the lengthened wheelbase means that there is capacity for nine passengers.

For many years UAZ produced Russia's only minibus. The chassis was the same as the UAZ 469's and the front grille gives it a rather doleful appearance. Its looks are not helped by a wheelbase which appears to be too short for the body. Variants include a crew cab pickup, van and standard pickup.

This new model is designed to replace the UAZ 469, but doubtlessly the two will be produced at the same time. This SUV has reasonably modern looks, and as well as power from the standard Volga engine, a fuel-injected Volga can be specified or indeed a VM diesel. Sales officially started in 1997 but would appear to be very limited indeed. There were plans to import the UAZ into Britain, powered by a Peugeot XUD diesel, but the idea (announced in a four-wheel drive magazine in 1994) has come to nothing.

ZIL produced its first vehicle in 1916, starting life as the Amo lorry works. In 1933 the name was changed to ZIS (Stalin factory). Car production started in Leningrad in 1933 with a boxy seven-seater saloon called the L1, based on the 1931 American Buick 90. The power unit was a 5650cc ohv with cast-iron pistons, a down draught Marvel carburettor and a very leisurely compression ratio of 4.4:1. The engine produced 105bhp.

In 1933 the Leningrad factory which produced the car transferred the presses and so forth to Moscow, to enable it to concentrate on other industrial kit. A group of designers under engineer Fachinski were put in charge, and the result was the ZIS 101 of 1936. The 101 Sport shown here was built in 1939. It could top over 100mph with its 8-cylinder 6060cc engine.

The ZIS had a 11ft 10in wheelbase, was 18ft 6½in long and weighed 5,500lb. Power was an 8-cylinder engine with a double-barrelled carburettor; a thermostat controlled cooling system and synchro gearbox. Other modern design touches included double acting shock absorbers, a heater and radio. The engine was bored out to 7570cc, giving a top speed of 68mph. The twin choke American Marvel carburettor was replaced by a domestic unit and power moved up to 110bhp with a compression ratio of 5.5:1. With the ZIS 101a the compression ratio rose to 6.4:1, but sadly the Russian petrol was not of sufficient quality for a further ten years. A few open-bodied ZIS 102s were constructed before the war, when production was converted to trucks.

1945 saw the launch of the first post-war ZIS, namely the ZIS 110, and production started in August only eleven months after the prototypes got government blessing.

Was the ZIS 110 a Packard rip off? There was no question that the ZIS looks similar to the 1942 Packard 180, and there have been articles stating that the body dies were removed from the USA. Every surface of the ZIS body is different, however, and it could be argued that the series 67 Cadillac had a stronger influence on the ZIS design.

The ZIS 110 remained in production until 1958, and the 6-litre engine produced 140bhp at 3600rpm, with a top speed of 87mph. The new model boasted hydraulic valves, independent front suspension and electric windows. Given the great length of the ZIS, a number were made into ambulances with no major body modifications needed: the stretcher was simply loaded through the boot. Because of a shortage of minibuses, the big ZIS was used as long distance transport, and a limited number of open-top models were produced. From 1966 the ZIS 110 became known as the ZIL 110 – when Stalin fell out of favour.

The ZIS 110 cabrio pictured was used by the Polish government before being imported into the USA, where it was photographed at the Greenwich Concours d'Elegance on Long Island.

ZIL does not stand for 'Factory named for Lenin' but rather Ligachev. In the mid-1950s the design studios of ZIL were starting to prepare a more modern design, the first prototype being known as the Moskva. The engine was uprated to 182bhp V8, rather than a straight 8, and an automatic gearbox was standard. For some reason the Moskva did not meet with government approval and was shelved. Three new American cars were purchased for study, including two modern Packards, the Caribbean and the Patrician. When the first ZIL 111 prototype appeared in 1958 the styling influences could clearly be seen.

The V8 engine had wedge-shaped combustion chambers and a cast-iron block. The 5980cc engine produced 200bhp and the auto transmission was similar to the American Powerflite: control was by push buttons on the fascia. The ZIL 111 remained in production for ten years, but numbers were more restricted than earlier models. This was because of development of minibuses and another large car, the Chaika, by GAZ. The 111 was improved on a number of occasions. The 111A was a stretched version, over 20ft long and with a weight of 6,205lb: the top speed was claimed to be identical to the earlier models, namely 106mph. The car got Russia's first air-conditioning unit. Also produced was the 111V, which had an open body and an electrically operated hood.

The fascia of the ZIL 111.

The ZIL 111 enjoyed a reskin in 1962 with quad lights, pictured here.

The ZIL 112 Cyklon really set Soviet pulses racing. This wild experiment with its cyclops eye was a one-off but achieved a very respectable mean of 109.2mph over 500km (318 miles).

The next model was the ZIL 114, designed by the deputy chief engineer, V. Rodionov. This gargantuan machine was even heavier at 6,720lb, despite having an aluminium rather than a cast-iron engine. The single four-barrelled carburettor produced 300bhp with a top speed of 118mph. The square looks were up to date with American machinery of the period and special fuel was required for the 9.5:1 compression ratio, although at least the tank held over 26 gallons. The official fuel consumption was 14.9mpg at 50mph. There were changes to the suspension, with the front suspension now fitted with longitudinal torsion bars; at the rear the conventional live axle continued. Braking was by four servo-assisted ventilated discs.

As well as electric windows, Russian leaders now enjoyed electrically adjusted seats, air conditioning, vacuum door locks, adjustable steering wheel, stereo radio, foglamps and tinted glass. As with previous models, twin batteries were used. Each engine ran on a dynamometer for thirty-two hours after assembly, and then each car was driven for 250 miles to check for any imperfections.

It was felt that a five-seater model was also worth constructing, and by 1971 a number of five-seater saloon ZIL 117s had been built, 18ft 9in in length and weighing in at 6,350lb. Owing to the lighter weight and 300bhp engine, this was the first Soviet passenger car to pass the magic 200kph (125mph) barrier. A little later a convertible version of the ZIL 117 was produced; it was known as the ZIL 117v. ZIL planners started to think about replacing the 114/117 as early as 1974, and the new model was, for some curious reason, known as the ZIL 4014. Production (if one can call it that) started in November 1978. As usual seven seats were offered, with two of the tip-up variety. Once again, there was a glass partition, air-conditioning and electric windows. However, much work had been expended on the noise suppression side. The vented disc brakes on all four wheels could slow the ZIL from 50mph to 0 in 44 yards while, once again, two batteries were used of a colossal 160ah. The 7695cc V8 engine had a bore of 108mm and stroke of 105mm, and the output was quoted at 311bhp at 4,000 to 4,600 revs. The four-barrel k259 carburettor was used, and the three-speed auto gearbox was carried over. Unladen weight was 7,337lb. Once again a convertible version was announced: the ZIL 41043. A five-seater model was also produced, seen here surrounded by crowds at the Moscow Motor Show.

The ZIL factory today is in a sorry state, although at least not completely bankrupt as Moskvich is. ZIL trucks continue to sell in reasonable numbers, although the ZIL light van has not been quite the success the management hoped it would be, given that the Volga Gazelle is the market leader. The ZIL design bureau, with windows facing on to the street, continues to produce pretty models and pictures – although the staff are still sensitive about photography: in August 1999 they successfully prevented me from taking photographs of the latest clay mock-up through the glass. Incidentally the only picture on the wall of the design bureau was of the Rolls-Royce Silver Seraph.

ZIL exhibited at the Moscow Motor Show in 1999 and officially the long- and short-wheelbase models are available, although whether any ZILs have been built in the last few years is open to debate. The price tag is a jaw-dropping US$70,000 for the 41041, or US$200,000 for the long-wheelbase 41047.

The first car to roll off the Moskvich production line in Moscow was a copy of the Ford AA. This was in 1930. GAZ at Gorky was constructing passenger and commercial versions of the Ford A at the same time. As the Ford AA dated, Soviet engineers started to design a smaller people's car along the lines of those being constructed throughout Europe at the time. This is the first home-grown car, the KIM 1-11, which bore an uncanny similarity to the Opel Kadett. Rumours suggest it also owed a certain amount to the Fiat Topolino. These are unconfirmed, but given the impunity with which the Soviet Union 'borrowed' bright ideas from around the world, it is not beyond the bounds of possibility. Production ran from around 1939 to 1945. Fortuitously, the Opel Kadett factory found itself in the clutches of Soviet forces at the end of the Second World War, and so it was transported lock, stock and barrel to the USSR, where the good burghers of Moscow reassembled it to produce the Moskvich 400 (below). Moskvich means 'son of Moscow', which in may ways this car was not; none the less, it found itself the basis for fifty years of production in the city. During many of these years the Moskvich was the only car produced for the masses in Russia, the Volga being a much bigger car in a different league. The Moskvich factory was snappily entitled 'The Moscow Factory of Small Displacement Automobiles' – or, as a less flattering translation had it, 'The Moscow Factory of Low-Power Cars'. Its initials in Russian were MZMA.

As is usual, slow development took place, with the Moskvich 401 and 402 being generally similar cars, with new features but mostly interchangeable parts. They were of pre-war appearance with separate mudguards and incorporating faired-in headlamps. Estate car versions were worked upon: the 'woody' is shown here. Thanks to the forward thinking of the Opel factory, unitary construction was employed.

Sadly most early Moskys are used as chicken coops or storage sheds.

The Moskvich 407 was the next incarnation – and, strangely, preceded the Moskvich 403 by some years. It had an all new bodyshell of late-'50s fashion, without separate mudguards, but again many main units were interchangeable and the car still incorporated the slightly quaint underfloor master cylinders, whose pistons pushed toward the rear of the car. Chrome strips adorned the sides and two-tone versions were produced. The interior was, and still is, most stylish, trimmed in patterned cloth with much attention to detail.

Not only was an estate version produced, along with a van, but the bodywork was grafted on to a teetering GAZ 69 four-wheel drive chassis, to produce a forerunner of such cars as the Subaru, where smart civilian bodywork is teamed with all-terrain capabilities. This car had all the ground clearance of the GAZ, and though its appearance is rather incongruous, it surely ranks as a classic of its time.

This racing car was produced on a 407 chassis in 1959.

Further work produced the Moskvich 403, which was really an old bodyshell on newer mechanicals. It is typical of Soviet circumstances that insufficient development funds were available to do the body too; but it is as well to remember that slow development such as this allows designs to be proven before they are used. The Moskvich 408 (pictured in a campsite in Warsaw) gave us the Moskvich body with which we are probably most familiar. It did service from 1965 to 1976, and then, with an occasional nip and a tuck, until about 1990. The gearbox and back axle were broadly similar to everything that had gone before. The engine was likewise – a 1400cc ohv unit of very low compression that ran on 76 octane fuel, which certainly smells more like paraffin than petrol. It was very user-serviceable, which was as well, given that in most parts of the Soviet Union garage service was a rarity. The air filter was of the oil-bath type; the oil filter was in two stages, with a washable coarse and fine filtration system. Tappets were easily adjustable through removable sections of the rocker cover, and on the high mounted distributor was an octane selector, which could be adjusted without ever getting one's clothes dirty.

The 408 bodyshell was a little larger than that of its predecessor and was of wider, lower appearance, as was the fashion at the time. Gone, sadly, were the smart cloth interiors, to be replaced with new, advanced PVC. New, though rather biodegradable, plastics formed many of the interior fittings, and the body-painted metal dash with padded trimmings boasted full instrumentation in a delightfully gaudy, pseudo art deco one-piece unit. Strangely, most of the instruments were of the slow-to-react thermal type, although the ammeter was a proper moving iron type and the petrol gauge a quality moving coil galvanometer. Thus, whilst it took all day to register a change in oil pressure, the Moskvich could record every slop and splash of petrol up the sides of the tank. The gauges led the tester of *Autocar* in 1967 to record that a journey in the Moskvich was 'rather like entering the Bermuda Triangle', as the gauges wandered to and fro in a world of their own. Still, at least it *had* instruments, unlike many Western cars of the time, which would indicate critical temperature only by violent pinking and a refusal to stop, even when switched off, and critical oil pressure only by a refusal to go when switched on. Like so many Eastern Bloc cars of its era, this Moskvich was a really quite advanced people's car, somewhat spoiled by the fact that the people found it extraordinarily difficult to buy one. As the 408 was coming on stream, the planners in the Kremlin took note of this fact and commissioned a second factory in the Udmurt SSR. They also started talks about the possibility of a joint venture with Fiat to produce copies of one of its current cars. The latter line of enquiry yielded the Lada, the former the less well-known Izh.

The early Izh cars were simply Moskviches sent, in kit form, to Udmurtia and assembled there, but as time went by presses and foundry installations were added to increase production. Eventually Izh was a virtually self-contained production facility. Izh is one of the oldest industrial concerns in the former USSR, having been founded in 1804. Its foundries and machine shops had for many years been turning out armaments and all manner of machinery, and in more recent times had diversified into motorcycle production. Its current two-wheeled product, the Planeta, is a 350cc twin which can be bought with extensive modifications, including an extended tricycle chassis with pickup and flatbed options. Surely the most bizarre device is a small fire appliance with a second 350cc engine on the back, used as a water pump. The Izh concern built only saloon cars at first, but in the very early '70s produced a Moskvich with a box van body on the back, predating the Muscovite attempt at a similar vehicle. A pickup followed. In 1972 production of the popular Izh Kombi started. This was a hatchback Moskvich, with semi-fastback styling. Its only disadvantage was a high loading sill, necessary for greater structural rigidity and crash resistance. In its time it was a smart vehicle and a fair compromise between saloon and estate, as are all hatchbacks.

As local content increased, the cars became easily distinguishable from their cousins in Moscow by, at first, even more lavish front chromework, and then by a total absence of it. A major order was to supply vans for the Russian Post Office: an example in Moscow is shown here. Other Izh cars, particularly the Kombi, found their way to Eastern Europe. Up until 1984 this was the only Eastern European hatchback car, other than the Zastava from Yugoslavia. It is interesting that this was also a saloon car, the Fiat 128, turned into a smart hatchback by its enterprising producers.

After this promising burst of activity, the funds ran out for further major developments at Izh, and basically the same cars were then produced until around 1990, when a new product, the Orbita hatchback (shown left), made its debut. This slightly dumpy hatch has Moskvich running gear and is manufactured at Ufa. It is available as a pickup truck or box van in addition. Although production of the Orbita is limited, the Izh concern continues successful manufacture of motorcycles, lathes, machinery, Kalashnikov machine guns and a huge number of assorted rifles and other armaments. It also builds, very successfully, Skoda Felicia cars from kits. Hyundai became involved with the factory, as did

The Izh Orbita four-wheel drive was launched in 2000, with the option of a 2-litre Hyundai engine.

The Orbita four-wheel drive estate remains in prototype form, and is pictured here at the factory showroom – where photography is not permitted!

Meanwhile developments were creeping along back at MZMA in Moscow. Roughly coincident with a change of factory name, a new engine was developed. In 1968 the Moscow Factory of Small Displacement Automobiles renamed itself as the amazingly catchy 'Automobile Factory of the Lenin Komsomol' or 'Lenin Youth' (AZLK). At about this time the Moskvich 408 became the Moskvich 412. Two reasons lay behind this change in designation, for the bodywork and trim of the cars were almost identical. Firstly, and most importantly, there was a second engine option. The incredibly heavy and lethargic car was transformed with the option of an all-alloy 1500cc unit of very high quality. It owed its basic form to the BMW 1500 unit, a slant four ohc creation of 80bhp (SAE). This reduced the weight of the car to only(!) a ton and improved its performance to compete in export markets with such cars as the Ford Cortina 1600. The bore of the engine was identical, the stroke 1mm shorter and the distributor was moved from the rear of the camshaft (for which the engine bay of the Moskvich was not long enough and that of the BMW only just so) to the oil pump drive at the front of the motor. The main change was of course to alloy block construction. None the less, the external appearance of the engine is almost identical, and the cylinder head virtually an identical copy. Many parts inside the engine are said to be interchangeable. The exhaust manifold is the same performance design as its Bavarian cousin and it gained a twin choke carburettor of incredible complexity which was, at least notionally, superior to the single choke Solex fitted to many BMWs. In practice, however, its complexity was often its undoing, and it is not uncommon to find frequent grumbles about flat spots on many cars. The new engine had to run on quality petrol of at least a two-star rating, owing to its higher compression ratio, but extra thick head gaskets were often contrived to allow very cheap 76 octane to go in. This was an especial advantage if the owner was friendly with a taxi driver or the pilot of a petrol-engined ZIL lorry.

The second reason for the change of model number is this. The second number of the model designation denoted the type of bodywork: 0 was a saloon car; 1 a special vehicle, 2 an estate car and 3 a commercial vehicle. As the Moskvich became widely exported, Peugeot pointed out that it held the international copyright on all three figure numbers with a zero in the middle. Hence the change from 408 to 412. Updated estate cars simply went from 426 to 427 (a somewhat worse for wear example shown here) and vans from 433 to 434. The 'special vehicle' designation was lost. Little matter, as the whole of the Soviet vehicle industry was soon after made to conform to a standard designation code of four numbers. Thus the next Moskvich to evolve was the 2138 range. Moskviches slowly lost the charms and idiosyncrasies of early 408s, as such outdated features as all-metal dashboards, art deco instrumentation and an amazingly dangerous pointed metal handbrake pull situated in front of the driver's right kneecap slowly disappeared. Many parts became supplanted with standard Lada items to reduce costs and make availability of spare parts for all cars easier.

The Moskvich 2140 of 1976 gained more Lada parts, a squarer, de-finned bodyshell and plastic bumpers. This was primarily an export model, and few are in evidence in Russia today. More common is the 2138 which retained certain external features of the 412, including elaborate chromed sidelight/indicator assemblies and metal bumpers. The oval Wartburg headlamps from the GDR, first introduced on the 412, continued. This car saw out the end of the model range. In the late '70s and early '80s the engineers and designers at Moskvich became desperate for a new model of car. The Lada was such a popular vehicle, with a generally more modern design, better carburation and more elegant interior and exterior, that even in the car-hungry USSR people were preferring to pass on a Moskvich and wait for a Lada to be allocated for them to buy. Furthermore the Moskvich, which as a genuinely good car in the early '70s sold well in export markets, was barely saleable to hard currency Western nations by the end of that decade.

Prototypes were developed, some based on the Moskvich 2140, others of seemingly original design. The car eventually settled on, however, took its bodyshell from the 1976 Car of the Year, the Chrysler Alpine. This design, fortunately, was updated by the Soviets, losing its slightly dumpy rear quarters: both front and back are seen in the picture above. Mechanically, it had a Moskvich alloy engine, longitudinally mounted, driving the front wheels. There was also an option to buy one with a Lada 1600 engine. Its extra performance must have made up for its extra weight, but as the gain can only have been marginal, perhaps this was part of a plan (which never reached fruition) to standardise engines between the two cars. Unlike the Lada, the Moskvich was never available with small engines. Rear-wheel drive was employed on commercial variants. One assumes these used the drive units of the Moskvich 412/2140.

The new Moskvich 2141 Aleko appeared in 1986, but the old saloon cars, which traced their ancestry back to the 408 of the 1960s, did not cease production until 1990. By this time, placed alongside a Lada Samara, the car was a bit of a relic. The Aleko sold reasonably well, and even reopened the export of left-hand drive cars to Western Europe, where it was often equipped by the factory with a Ford diesel engine for greater economy. However, it was not the success that the high performance 412 had been in its day. It is a worthy car, modern in appearance and, with a styling update, continues in production to this day. Sadly, it too has become something of a relic, with poor sales and poorer build quality. Frequent lengthy pauses in production occurred during the chaos of the early '90s, with production lines of half-built cars standing idle, only restarted occasionally thanks to orders for taxis and municipal vehicles from the Moscow city authorities. The 1997/98 update of the cars started things rolling again, but the factory remains a basket case. Cash, meanwhile, has been poured into pointless tooling to produce long wheelbase and even 'luxury' variants. The current 2142 model sells as the Sviytagor, the Kalita and the Prince Vladimir. It is hard to see who will lay down good money for a luxury variant.

The two-door Sport is available with Lada and Renault powerplants, but production is unlikely to have reached double figures.

During my visit to the car plants of Russia during 1999, the factory yard was full of half-completed vehicles. Money was simply not available to provide such parts as batteries, dashboards, steering wheels, instruments, headlamps or even tyres. Cars sat perched on piles of half bricks, awaiting completion. This would be bad enough, but, almost criminally, often they had been dumped with windows and doors open, allowing rain to soak carpets and soundproofing and to ruin seats. It is not difficult to see why Russians are reluctant to buy them. No Muscovite searching for a well-built car would go near them, having seen this. Marketing initiatives push a Renault 2-litre engine, a 1.7 successor to the alloy 1.5 litre Moskvich engine and the Lada 1600 motor. Long-distance proving runs are publicised – Moscow–Omsk–Vladivostok in 1999, and a seventieth anniversary run leaving Moscow and calling at Arkhangelsk and Murmansk in 2000. But the fact remains that with cars half built, and finished off piecemeal, build quality must be dreadful.

What is the future for Moskvich? It is hard for anyone to say. The Moscow City Authorities are the majority shareholders in Moskvich and ZIL. Production facilities are being shared and municipal orders keep the factory ticking over. An unpopular and outdated range (the Samara is a much better buy, even if much more expensive) needs urgent replacement, but there is simply no money in the kitty. Renault has been making interested noises for some years, but the time is never right, it seems. Even a plan to assemble kit-form Méganes seems to have fallen through. Sadly, in the long term, the Moskvich plant, its huge social facilities and football stadium, along with other benefits for its dwindling workforce, seem to have little future.

From the early '50s small numbers of Moskviches were occasionally imported into the UK. It is said that a few 402s came in, but an attempt to sell the 403 was thwarted when it was found to infringe a number of other manufacturers' patents. The Moskvich 408 was imported and sold in small numbers for the price of a Ford Cortina. The sales breakthrough came when the concession for importing was taken over from Russian Cars Ltd of Byfleet in Surrey by Satra Motors (Soviet American Trading) and an import centre built at Carnaby, near Bridlington. It is said that a field of long-unsold 412s was cleared from Byfleet and taken to Carnaby. The cars had sat in the field so long that many had to be de-rusted and resprayed before they could be sold on. Satra realised that the car could be sold at a profit, even if its price were reduced to that of a Mini, and that is precisely what they did. Between 1970 and 1973 sales soared: the picture above is from the British brochure of the time. The Moskvich, unfortunately, became a casualty of the Cold War. The Consumers' Association brought out a report damning the Moskvich on grounds of safety. Breaking from its usual practice of first allowing the manufacturer to see and comment upon the report before its release, the CA gave it to the press on the eve of the 1973 Earls Court Motor Show. The shock waves rang out. The BBC programme *Nationwide* ran a feature, the motoring press went wild and sales of the car died overnight. The last models were sold in 1976. In 1974 Satra Motors started to sell the Lada, which, owing to its Fiat origins, was less open to such criticisms of its design. Although the CA report made some valid points, for instance on the subject of some dangerous protrusions in the passenger cabin, other items on the list were simply down to the test car needing adjustment to the steering and brakes. Further items were no more than the product of a fertile imagination. For instance, if the wheel nuts are not tightly done up the wheel may fall off, shearing its studs. Well, who would have thought it? Given that the Moskvich had five stud hubs instead of the three or four of other cars, this made even less sense. In Britain we never saw the Moskvich again. In the mid-'80s a small number of Alekos were imported and put on long term test by Satra, but this came to nought. Izh even sent a Kombi for evaluation, but this was never even considered and was sadly put in the crusher quite recently by Satra's successor, Motor Vehicle Imports.

My own Moskvich is an early '70s 427 estate. As it will not be roadworthy until after this book goes to press, let us sit in it and peruse its virtues and vices, then switch to a 1978 2140 saloon for a test run. The first thing to notice is that the doors open and shut with a completely un-Soviet efficiency. The smoothness of operation is more akin to German cars and gives a good impression of the attention to detail in all mechanical units of the car. This car was certainly built to a specification, rather than a price. Not content with building a close copy of a BMW 1500 motor, the designers at Moskvich took it upon themselves to build in special features. The original Moskvich 408 was a very heavy car; when the 1500 engine was designed, they elected to reduce the kerb weight to a mere ton by completely re-engineering the unit in a high quality, corrosion-resistant aluminium alloy. The gearbox remains an anachronism. Dwarfed by a giant alloy bellhousing, it is the size of a loaf of bread and is of cast-iron construction. The front suspension is double wishbone with swivel joints, not kingpins. This is all bolted to a heavy forged crossmember. The rear suspension is on prehistoric leaf springs with heavy duty shock absorbers, which are as good as the day they were made, thirty years ago. Indeed, the attention to durability and easy repair runs throughout the car, with all sorts of units having shims to adjust out wear and being fully 'tunable' with screws and bolts for precise setting. Definitely the only part of the car to have escaped the design department is the interior. Even at the time it was a bit of a horror. Although the seats fold flat to make sleeping accommodation, they are not reclinable in normal use. They are, fortunately, comfortable and supportive. The dashboard, although described as 'padded', is in fact steel with occasional padded strips and the glove-box lid, though spring loaded, threatens to leap open in an accident and decapitate any passenger foolish enough to forego the seatbelt. The handbrake is a pull-handle of cast aluminium, ominously located near the driver's right knee, and a host of protuberances demand careful driving to avoid the painful consequences of an accident. Still, the whole interior is delightfully quaint and thus infinitely preferable to the offerings of 1973 onwards, which got a between-the-seats handbrake, fully padded dash and depressingly '70s instruments. Our older instruments beam forth in a chromed art deco display of gaudiness; a single steel stalk controls indicators and acres of Bakelite gleam around the shiny black steel dash like the trimmings of a valve radio. This car has class in spadeloads. Controls are scattered around, on and under the dash for convenience of manufacture, rather than any convenience of the operator, and a huge cast aluminium heater sits atop the transmission tunnel like the pressure vessel of a nuclear reactor. Unusually for its time, body coloured metal is little seen in the interior, although the metal back of the estate rear seat is finished as per the exterior. Does this sound negative? I hope not – this is the delightful essence of the Mosky, some compensation for the neo-gothic exterior with grinning chrome grille wedged between alien oval headlamps bought in from the GDR after the car had been designed for round ones from the USSR. This, and the engine, the major selling point of the car in the West, give the Moskvich its cachet and desirability.

We will now potter across to the 1978 saloon belonging to Tony Johnson and go for a spin. The engine growls to life and revs with increasing roar, rather than increasing pitch. The controls are surprisingly light and the gearshift quite precise, in spite of its being worked through a linkage. It sets off with stately progress, torque is here in abundance. Most noticeable now is the gearbox. The penalty for having a gearbox smaller than a glove-box is a loud and insistent whine in all indirect gears. Owners of Austin 1100s, rejoice! Reaching top gear, little thought is now paid to the engine. Although the car does not feel fast, this is because of the very high driving position creating the illusion of measured progress – in fact, the car does not want for more power at any time. The steering, though of much the same type as a Lada's, is as precise and light as a rack and pinion setup and the brakes (front discs on this model) are firm and responsive. No wonder this car was a saloon car racing champion of its day. Bends are taken with ease: the body rolls magnificently, owing to its high centre of gravity, but the East German cross-ply remoulds hang on with grim determination and the motor never protests.

Is this quirky antique the zenith of early-'70s' motoring? No, not by a long chalk. But you would have to go a long way to find a car so endearing.

The history of Lada goes back forty years. In the 1960s the mainstays of Soviet passenger car construction were threefold. The small motor car sector was catered for by the Ukrainian Zaporozhets, a rear-engined car with a 1200cc air-cooled V4 motor. This was ideally suited to cold climates and unmade roads, its configuration giving maximum traction where little grip was available. The medium sector was covered by the Moskvich with a choice of 1400cc low compression engine, capable of running on the 76 octane petrol (which was often the only fuel available in rural areas) or a higher compression 93 octane 1500cc unit derived from the BMW 1500. Large passenger cars were provided by the GAZ works at Gorky (Nizhny Novgorod) whose Volga range had been in production since the mid-'50s. All these were tough, proven designs. It became evident that a great increase in production was necessary in the medium car sector. To some extent this was provided by the production of Moskvich cars outside Moscow at Izhevsk in the Udmurt SSR, which started in 1968 and increased its output as metal stamping, engine and other facilities were added at the Izhevsk works, on the banks of the river Izh. Some years before this, though, serious thought was given to the fastest and most efficient way of developing both a new production facility and a new model at the same time. It was decided that outside help should be sought and, after much consideration, a deal was brokered with Fiat. Crucial to these negotiations was the leader of the Italian Communist party, Togliatti.

In 1966 preparations to build a new production facility were instigated at a site near the river Volga in the Samara region, some 500 miles south-east of Moscow. This was to involve not only a factory but also a new town, to be named Togliatti in honour of the man who had been crucial to the whole deal. The VAZ factory was to be unique in Soviet motor vehicle production in that most of the parts for the new car would be made on site. Only a few items would be brought in from outside, notably certain ignition equipment, lighting units and window glass, which were to come from the existing Bor works at Gorky. Production began in 1969, when the works produced some 30,000 cars, but this was to be a fraction of its intended future output. The car model produced – the Zhiguli – was a derivative of the Fiat 124, a generally modern design which retained the traditional features of a live axle, rear-wheel drive front-engined layout.

The Fiat was not suitable for direct copying, however. Conditions in the Soviet Union were, and indeed are, very harsh for motor vehicles. The temperature might vary between -60°C in the winter and 50°C in the summer. In addition, main roads are very poor because of heavy vehicles and extreme frost-shattering in winter. Side and rural roads are frequently no more than dirt or mud tracks. In remote areas, there are no roads – merely directions in which one drives – and it is frequently the case that in the dry season cars are driven along river beds from one place to the next. Thus, the 124 with its thin metalwork and low, semi-sporting suspension needed modification. In the tradition of Soviet engineering, the car was first redesigned to use metal 30 per cent thicker than the original Fiat. Modifications were made to the suspension and steering to reinforce these units and the ride height was raised to give a minimum ground clearance fully laden of over 6in. The front disc brakes were retained and modified to use two pot calipers, and the rear discs were replaced with drum brakes to cope better with rear axles constantly run through mud and snow. As usual, developments were made to a specification rather than a price, and so enormous 10in drums were specified. To reduce unsprung weight these were made of aluminium alloy with steel friction linings. In addition, exhaust systems were reinforced and other parts including the floorpan were carefully strengthened.

In the engine department further changes took place. The engine unit was increased in capacity to 1198cc and its design altered from ohv to ohc operation. This produced a unit of oversquare dimensions (76 × 66mm) which would give a respectable 62bhp at 5600rpm and 64lb/ft of torque at 3000rpm. It was of iron block/alloy head construction, had five main bearings and a coolant capacity of 16.8 pints which was even further expanded (to take account of desert conditions) from 1976. It is acknowledged, particularly in its later bored-out versions, as the finest asset of the car, by virtue of its extreme ruggedness and load capacity. These reinforcements, however, led to a considerable weight penalty and the new VAZ 2101 car came in at 2,105lb dry mass. Although the economy penalty was not too severe (40mpg being achievable on a run), steering, being of the tough but unrefined worm and .roller variety, became somewhat heavy. Early models were notable for having little self-centring action. The gearbox was both tough and reliable and the rear axle famously unbreakable, even if a little noisy if improperly adjusted at the works. As the Togliatti works grew, so did its model range. VAZ 2102, the Fiat 124 estate car Sovietised, appeared in 1971.

In 1973 VAZ 2103 was introduced. This boasted an engine with a longer stroke, taking it to 1452cc, and many minor improvements, the most practical of which were twin headlamps, further improving the excellent lighting performance, and the addition of a vacuum servo to increase braking efficiency. Full instrumentation with tachometer and oil pressure gauge was added, and all body panels and interior appointments were modified to distinguish the car from its humbler brethren. 1974 saw the VAZ 2101, a base model with some trim improvements and either a 2101 engine or the same engine bored out to 1298cc capacity (67bhp). The VAZ 2103 was offered with a choice of all three engines. In 1976 the VAZ 2106 superseded the 2103. It offered a slightly updated style with improved rear lights and was available with a choice of 1298cc, 1452cc or new (model 2106) 1570cc 80bhp engines. Little development now took place, efforts being concentrated on minor improvements in design. Exports had started early on.

As the '70s drew to a close the designers at Togliatti were working on a reskin of the faithful Zhiguli, and in 1980 model 2105 appeared, followed by model 2107. Mechanically, improvements were minor. Improved carburettors and ignition systems (at last, a vacuum advance in place of a manual octane selector!), and better steering design comprised most of the mechanical updates. The body was squared off, in the fashion of the late '70s and pencil-thin front wings were a consequence of wider front track, which used the same parts on wider suspension pillar spacings. In addition, a sweet belt-drive engine of 1298cc was produced (eventually) to supersede the chain-driven 1298 of old. Large, rectangular headlamp/indicator one piece units clad the front, with wash/wipe a standard fitment, and the rear was dominated by huge and very efficient all-inclusive light clusters. Model 2107 was distinguished by a large chrome grille and different rear lights, all of which aped the new ZIL limousine of that year. In addition a much improved interior and better instrumentation were specified, which brought the car right up to date.

Less stylish was the pickup version, with welded-up rear doors.

After the launch of the 2105 and 2107 the other models continued in production. Meanwhile plans were afoot for a completely new design. Two avenues were explored: a new Soviet design and another bought-in Western product. Whilst both were eventually adopted, the one which came to fruition first was the VAZ 2108: to the Soviets, the Zhiguli Sputnik; to Western markets, the Lada Samara. Principal in its design was the Porsche company, which supervised the whole operation and applied its expertise most notably to the engine. British design found its way into the transmission and brakes. The car bore no resemblance to the former models; indeed, further than nuts and bolts, no parts were carried over. It was a front-wheel drive three-door hatchback with Macpherson strut front suspension and a torsion beam rear axle, in accordance with the practices of the day. The engine lay transversely in the front of the car, coupled to the wheels by a five-speed gearbox and differential unit. In the view of the Porsche company, its greatest achievement was making 'the first Western vehicle capable of coping with Russian conditions'. The 2108 went into mass production in 1984, but the five-door 2109 did not appear until 1988. 1990 saw the model 21099 saloon car. Porsche-modelled engines came in 1099cc, 1288cc and 1499cc variants, the 1099cc appearing only in the three-door car. As if to emphasise the slow evolution of the VAZ product range, 1984 also heralded the estate car VAZ 2104, based on the now four-year-old 2105 saloon. Slowly the 2101, which first saw light of day in 1969, was phased out of production. The 2102 estate car yielded a year or two later. The four headlamp 2106 model continued, however.

The strengthening needed on a Samara convertible resulted in a 'shed on wheels' effect.

The facelifted Samara, the 2115, has recently been introduced, with a top speed of 115mph from its 1500cc engine. The major change is its new nose.

In 1988 another new model, which is not widely known in the West, appeared. The VAZ 1111 Oka was a small town car, about the size of the little Japanese Daihatsus and Suzukis of the day. It continues in production to date. It has a 2-cylinder engine of 649cc, which is in fact half of a Samara 1300 power unit. It is produced in three centres: the main Togliatti works, the KamAZ truck works at Kama, and the SeAZ factory at Serpukhov. The Oka has a performance disadvantage, weighing 1,430lb but only having a 29bhp engine, leading to what one may politely call leisurely acceleration; but as a town car, it excels. Its worst consumption is 40mpg in city traffic and its turning circle is just 15ft – or, put in perspective, half that of a Nissan Micra or any other rival. The Oka is a steady seller, if not a runaway success, but its latest Togliatti incarnation (the Lux) ditches the many interior fittings from the VAZ 2101 (which had been out of production for a good four years before the Oka's appearance), and replaces them with those of the VAZ 2106: the round dials of water, petrol and oil gauges complement the speedometer, but one is at least spared the indignity of a rev counter in such a pedestrian performer. In fairness, its 2-cylinder engine cannot fail to be a zingy unit, free from half the moving parts of a full-sized design, and the thirty seconds it takes to wind itself up to 60mph is explained by the fact that 65mph is top whack. But the design parameters of a city car mean that its economy, turning circle and consequent ease of parking its 10ft 6in body make it ideal for its purpose. The KamAZ works presented a body restyle with 'retro' round headlights at the 1999 Moscow Motor Show (left). The cars built at Serpukhov are all adapted for the disabled. AvtoVAZ also promotes the OKA-elektro, which drives rather better than the real thing, owing to extra weight placating the springs and extra performance in low speed acceleration. This is a very sensible conversion, electric cars being ideal for city driving conditions.

As mentioned earlier, the front-wheel drive car project which yielded the Samara also had a home design option. This was worked upon by the very talented development team at Togliatti for many years. It is typical of the Soviet motor industry that in spite of some of the world's best trained designers working in dedicated facilities, lack of investment precluded the production of most of the fruits of their labours. Even successful ventures were outdated by the time production lines were set up. Upon completion of the VAZ 2105/2107 cars in 1980, work on the new Soviet/Russian VAZ 2110 began, and the new car finally hit the streets in 1997, as the Lada 110. Prototypes had been tested from around 1994, and by this late stage were using many tried and tested components of the Samara, most notably the engines. The Lada 110 is bigger than the Samara, though not by a great margin. It is available as a saloon (110), estate car (111) and, hatchback (112), which is notable only by its absence from Russian streets. The 110 and 111, however, are numerous in the cities and appear to be selling well, despite costing twice as much as the base model Zhiguli – VAZ 2105.

The Lada 110 is a well-proportioned car whose large rounded nose and high tail hint at late-'80s Audis. As a piece of design it hangs together well, let down only by 13in wheels which sit, like castors, dwarfed by the high slab sides of the vehicle. It is possible to buy cars with 14in wheels, which seem somewhat better proportioned. Inside, there is a large, sweeping dashboard and curvaceous instrument binnacle with arrays of switches spreading right the way down the centre console. On the outside, body-coloured plastic swoops down from the bonnet and boot, although the swoop is interrupted by the panel gaps which are such a trademark of Eastern Bloc construction. The thin headlamps have been further developed with the adoption of behind-the-glass prefocus lenses; air conditioning and power steering complete the impression of modernity.

During the latter days of the development of the 2110, AvtoVAZ became involved in cooperation with General Motors. The two companies have been something of a *bête noir* to each other during their years together. GM, entranced by VAZ's enormous market share and burgeoning R&D expertise, has kept a safe distance during the economic chaos of the early '90s and AvtoVAZ's huge debt to the Russian government. VAZ, likewise, has been suspicious of GM's intentions with regard to the current focus on a wide range of popular but outdated models, and its likely plans to streamline Togliatti's 125,000-strong workforce. None the less, GM influence has crept into the Lada 110, most notably in the 2-litre GTi model, seen here with a poor-taste bodykit. This is Russia's first production sports car available to the mass market and uses an Opel 16-valve engine first seen in the Opel Kadett/Vauxhall Astra GTE. GM fuel injection systems are also used on other models in the 110 range, and on some Samaras and Nivas.

But have AvtoVAZ left behind the days of the infinitely repairable car designed for the home mechanic? Service facilities outside towns are a rarity, and a breakdown in temperatures of -40C spells death from hypothermia unless a roadside repair can quickly be made. Moreover, Russians love their Zhigulis because they have been familiar with their every idiosyncrasy for some thirty years, and because a part from a long dead Zhiguli can be bolted on to an ailing new one to effect an instant repair. This explains the continuing popularity of model 2106. This veritable anachronism is still in production. When, in about 1996 or 1997, Togliatti needed production space for its newer models, the tooling was removed and sent down the road to a new facility, RusLada at Syzran, where this car, now marketed as a 'classic in its own lifetime', is produced in ever-increasing numbers. Technical developments from later models are included, such as collapsible steering columns, sports seats, new carburettors and electronic ignition, but this car continues very much as a Russian Morris Minor, four headlamps and new Duralumin bumpers smiling out from the latest fashionable paint colours. Retro chic has arrived at Lada quite by accident. The car is, in terms of its appearance, the exact same as the 1976 model, minus a little more chromium plate each year, and the Russians love it for its combination of agricultural charm, unbreakability and sporting (and unburstable) 1600cc engine, which will push beyond 8000rpm (some 30 per cent beyond its notional safe limit) when the occasion demands.

The 2105/2107 range continues, with most sales seeming to be of the upmarket 2107, which is definitely built to a specification rather than a price. The 2104 estate car is enjoying something of a sales boom, being the only Russian estate car in production now, other than the cavernous and more expensive Volga. Pick-up variants of the 2104 are now produced, the first commercial variant of this car, and a normally aspirated 1500cc diesel engine is an option on this and the standard 2104. The performance of this low-capacity oil burner is as one would expect, and the press office once described the diesel 2104, without a hint of sarcasm, as 'for those who are not in a hurry – dacha owners, tourists and anglers'.

We have come thus far without a mention of Togliatti's most internationally acclaimed model, which has won its reputation as amongst the very best in its class, despite seldom competing in terms of luxury or top speed. It has been in production since 1978 with only one minor design update and is only just about to be superseded – though one supposes not replaced. The VAZ 2121 Niva is one of the few vehicles in the 4×4 class that can truly be said to have been designed simply for driving to inaccessible places. The Soviet Union and its successors, forming the world's largest land mass, have a unique environment. It is almost inconceivable to us in a small country such as Britain that one could travel for hundreds of miles without ever meeting a road, but in a country such as Russia or its satellites, this situation is commonplace. With the development of the countries' natural resources, ever more access is required to remote locations. The principal mode of transport in this outback is the UAZ, derived from the proven GAZ 69 army Jeep, but the enterprising development team at VAZ produced a unitary construction car using four-wheel drive in conjunction with the engine from the successful VAZ 2106. With a negligible front and rear overhang, some 9in of ground clearance fully laden and a passenger car interior largely derived from model 2106, the VAZ Niva was bound to be both a popular choice at home and a sales success abroad. Many of the interior fittings came from the VAZ 2106 and new suspension and drive units were designed specifically for the car. Forgings and castings replaced the pressed steel suspension components of traditional Lada cars and the short, rigid bodyshell took the place of the separate chassis usually employed on such vehicles.

The Niva was, in many respects, a modern vehicle for its time and in spite of other manufacturers' efforts in modernising and developing four-wheel drive passenger vehicles for general sale, it remains one of the best performing all terrain cars. Development of the car has recently been undertaken. The basic model has had its 1600cc engine bored out to 1700cc and interior fittings modernised, with many Samara parts replacing those of the ageing 2106 model. A full-length tailgate replaced the previous three-quarter length door – given the necessary torsional rigidity needed in such a vehicle, the removal of the tall rear sill was no mean feat. Long wheelbase variants in two- and four-door versions (above) have appeared in the last few years and a rear-wheel drive long-wheelbase van is in limited production with an electric motor. This is said to drive well, except in respect of its brakes, which lack engine-driven vacuum servo assistance. Thus braking is safe and adequate – a noted feature even of the early non-servo Zhigulis – but requires a very firm push of the pedal. Pickup and crewcab Nivas (below) have also joined the line-up.

Exhibited at the Moscow Motor Show of 1999 were the intended successors to the current Niva range: swooping curves, contoured lights and sweeping fascias (below) denote model 2123. Its engine has acquired fuel injection, which is tuned to increase torque at low speeds. Although looking very contemporary in its styling – which somehow is more convincing than that of the 110 saloon car – it none the less retains minimal overhangs and maximised ground clearance. Although prototypes have been around for a few years and pre-production models are widely publicised as part of the current model line-up, no 2123 cars are in evidence in Russia; a symbol of AvtoVAZ's continual fight to find money to put production lines together, and difficulty in finding space in a factory already running at full capacity and producing easily sold models of older designs.

Although Western manufacturers produce model ranges larger than those of AvtoVAZ, differences between models often comprise trim upgrades and fitting of extra equipment. It would be fair to say that although many VAZ cars are available in only one or two levels of specification, the number of separate models is, by most standards, immense. But our amble through the blossoming catalogue of Togliatti production would be incomplete without a look at some of the special vehicles produced at the plant, some in conjunction with outside firms (both Russian and Western). There are some vehicles here which would never be seen in the portfolio of a mainstream manufacturer in the West.

The two electric vehicles based on the Oka and Niva respectively use conventional lead-acid batteries and thyristor control. They are a result of almost twenty-five years of development, which has seen rather boxy light commercial vehicles built on existing running gear. The most numerous of these were based on the model 2102 estate car mechanicals. The bodies were always rough contrivances – styling was not the object and a small fleet was built in 1978 to complement a government initiative to set up children's driving schools. Needless to say, this did not go ahead and most vehicles finished up on factory duties. The Oka-Elektro was really the first successful marketing of this concept, and a great deal was learned by rallying these cars in international electric-traction events. The electric Niva went on sale a couple of years ago.

The Niva also holds the world altitude record – the only vehicle to reach over 18,500ft under its own power. A Lada team achieved this in the Chinese Himalayas.

The Niva has been adapted for other duties during its production run. Niva Force is an armoured version for bank delivery runs. The remarkable Niva Marsh, illustrated here, has an extended chassis carrying huge balloon tyres. This is intended for traversing wetlands. Not only do the tyres spread the load safely over a large area of ground; they also create enough buoyancy to allow the car to float in water. The rotating wheels then act as a means of waterborne propulsion.

The most bizarre Niva contrivances, however, hark back to the days when the Moscow Factory of Small Displacement Automobiles (MZMA) grafted Moskvich bodies on to teetering GAZ 69 Jeep chassis. The Lada Tarzan is a Samara body spliced with a Niva chassis and brought back into proportion by the use of great swathes of plastic air dams, side skirts and spoilers. A similar car has been producing using a Lada 111 bodyshell. The result is a car which, despite its stylish aspirations and beefy appearance, might struggle to find a market niche. One rather suspects it is a rally homologation exercise, as it is too tough to use as conventional transport, yet too smart to use as an off-road hack.

Lada Aero represents, like the Niva Marsh, the sort of vehicle that most manufacturers simply would not find sufficient market for. In Russia and its satellites, however, unique conditions exist which turn such a vehicle into an economic reality. This car has as its base a Lada Samara bodyshell. Into this is placed a 100bhp 1.9-litre petrol engine and a fifth wheel. This is housed in the boot. It is in fact an aircraft wheel, and it is connected to a hydraulic landing gear and braking system. Its purpose is to measure accurately the coefficient of braking on the sort of landing strips which are found in remote areas, servicing both outlying settlements and installations for oil, gas and other natural resource extraction.

Finally, competition cars are enjoying a comeback in the AvtoVAZ line-up. In former times tuned versions of Zhiguli cars took part in both track racing and off-road rallying events in the Soviet Union and its neighbour states. Though not competitive with Western entries, which had much larger engines, they none the less provided a useful publicity boost for a company dedicated to more humdrum products. In very recent years, in conjunction with other firms, competition cars are once more a source of pride for this expanding giant of the Russian motor industry. Other than the electric Oka, which was rallied in foreign events to gain much-needed test information, the most recent attempts are based around the 110 saloon car. One uses a GM fuel-injected 2-litre engine the other a Russian twin-chamber rotary engine of 1400cc capacity, producing around 200bhp. Whilst one assumes the former car cannot fail to be successful, it is more difficult to ascertain the success or otherwise of the rotary engine. Russian rotaries were at one time fitted to certain motorcycles, mostly for competition use. Whilst, given the experience of other conventional Russian engines, the quality of materials and engineering may not be in too much question, the actual tolerances that these motors are built to is a critical factor. Possibly this is less of a disadvantage in competition use where engines are rebuilt after each race; the famous NSU rotary débâcle only seemed to begin as engines exceeded the 10,000-mile mark.

Concept cars, such as the 1102 small car prototype of 1973 (above) and the Elf Elektro of 1999 (below) appear regularly at motor shows.

A new people carrier with rather awkward styling was introduced in 1999 based on the Niva floorpan. This example was photographed on a wet day in Samara.

VAZ 1119 Kalina is Lada's new compact car. Though versions have been released to the motoring press, no production date is yet in sight.

The first right-hand drive Lada to hit the shores of Great Britain arrived in 1969. It was romantically badged as a Soviet Star. Rumour has it that this car still exists in private ownership. A further couple of cars appeared in 1972, at least one in left-hand drive form. These cars were shown to the press and were available for reporters to test drive at special motor trade days. They were favourably received, being acknowledged as a much cheaper version of a well-known current Fiat design. Few really cheap cars were around at the time; the Skoda and Wartburg were established contenders with steady, if modest, sales. The existing Soviet import, the Moskvich, was enjoying meteoric sales success since Satra Motors had reduced its price tag from Cortina to Mini level. Though not much enamoured of some of its Eastern-bloc crudity, buyers were won over by its 1500cc engine and 90mph performance, which blew most Western competition of twice the price right out of the water. By 1974 Satra Motors had its first batch of VAZ 2101 and 2102 cars. Early models of car still bore the Cyrillic badge Zhiguli, but by 1975 smart enamel badges announced its export name, Lada. The Lada 1200 saloon and estate were marketed as tough cars, as indeed they were. Being so much based on the Fiat, no one could grumble about the quality of the design this time. Keen pricing attracted a few members of the buying public at first. Then the cars began to sell more and more, largely because of word-of-mouth recommendation by satisfied owners. Sales rose; the range expanded. The 1500 saloon (model 2103) arrived in 1975; by 1977 the 1300 (model 2101) had appeared and the 1600 (model 2106) hit British roads in 1978. The Niva came out at the same time – at first in left-hand drive only, then in British-converted optional right-hand drive, then finally in the early '80s in factory right-hand drive. This was an instant hit with farmers. Keen pricing and sincere marketing (they were sold for what they were – strong, reliable and cheap) increased sales steadily. The cheapest models were always the bestsellers.

In 1983 model 2105 appeared as the Lada Riva. Once buyers were satisfied that the car had not been cheapened in terms of its remarkable strength and quality, sales took another leap forward, and the introduction of the very highly specified 2107 and the estate saw the discontinuing of the earlier round headlight models. As the squared-off Riva appeared, so Western manufacturers once more left the Lada behind as the jelly-mould look came in. None the less, British buyers still liked the Lada for its traditional values of strength, reliability and owner servicing. By 1988, when the Lada had really become an anachronism, sales were peaking at 30,000 per annum. The Samara model appeared in late 1987, and gained in popularity as a modern offering with old-fashioned strengths.

Sales boomed until the introduction of catalyst legislation in 1992. Whilst catalyst Samaras seemed quite well sorted, Rivas never really adapted to the extra complication and quickly gained a poor reputation amongst long-time Lada owners. Carburettors were the main problem. Agents seemed unable to cure problems and sales took a nosedive. In reality, a strip-down and rebuild of the carb was all that was required, but by now Lada UK was also selling Malaysian Proton cars through its agents. Dealers seemed much more interested in this modern product. A combination of these factors and the now non-existent marketing of Lada cars in the UK culminated in a row between Lada UK and AvtoVAZ. In 1996 it was announced that whilst a spare parts contract had been renewed for a further five years, the supply of new vehicles was to stop. The Lada was officially finished in Britain.

Let us step into my 1988 Riva 1300SL. Officially, it is model 2107, so it is fronted by a large chrome grille. It is finished in surprisingly lustrous midnight blue, which still comes up like new after twelve years. It has now covered 120,000 miles. Sitting in its red velour high-back seats (which are generally assumed – wrongly – to be an after-market fitting) we gaze upon a large one-piece instrument panel housing speedometer, rev counter, oil, temperature and fuel gauges and a voltmeter. The design hangs together well and still looks fresh and clear, marked out in four colours, twenty years after its inception. The seats are comfortable and supportive and all the controls are to hand.

The engine fires first time, every time. The notchy gearbox, which once required a strong will (and arm) to engage first, now slides effortlessly into any gear – run in at last! The synchro is smooth and impossible to beat, however fast the gears are changed. The clutch is smooth but requires precise control: the strength of the design means there is little driveline cushioning from clumsy use. The steering is, famously, very heavy at a standstill and somewhat vague on the move. The throttle has inches of free play, owing to a necessary clearance of 2mm at the carburettor being magnified by the linkage.

So the controls are (gearbox excepted) nothing special. On the move, however, the car is a surprise. The 1300cc engine copes remarkably well with the 1 ton unladen weight and will propel it from 0–60mph in 14 seconds, cracking 93mph on the flat. The sheer weight of the car means that, with its long travel suspension, bumps are of little consequence, and on unsurfaced tracks the car can be hustled along with a minimum of jarring and discomfort. If the 67bhp of the engine needs to be used in anger, it will race well above its notional 6000rpm limit and on to 8000rpm without vibration – though I imagine a supply of conrods might be a wise investment for the rev-crazed pilot. In reality, the admirable torque means it is seldom necessary to drop out of fourth gear and never lower than third. It can be driven smoothly and quietly up to 60mph; the determined motorway cruiser should seek out a five-speed model for quiet high speed motoring, however. Even 90mph can be held with ease, and 40mpg can be returned on a run, 35mpg in normal use.

The headlamps are powerful and come with beam height adjusters and wash wipe – features usually found only on luxury cars of the era.

Although the Lada was often subject to music hall jokes about its reliability, this was generally only true of ill-maintained examples, of which, sadly, there are many. In the 100,000 miles I have driven this car, it has never failed to complete a journey. One dead battery and a faulty ignition connector are all that have ever troubled it. At 113,000 miles a piece broke off an exhaust valve. Upon opening up the engine, I was amazed to find no bore wear at all and wear of only a thou or so on the camshaft. I can only say that this car is of very high quality.

The main snag is that it is no more than an old Fiat in the body department and as such needs regular rust treatment. This is the car's sole weakness. Application of oil into the box sections is the cure. Given the number of rust traps, this needs to be done religiously each year.

Would I buy another? No. I have often considered it, but I just can't find any reason to get rid of this one. Only a catastrophic accident is likely to force its renewal.

The future seems bright for AvtoVAZ. Its troubles since the demise of the USSR stemmed mainly from typical post-Communist cashflow problems. Critically short supplies of some materials led to scarcity of certain parts and abysmal quality of others. A fine example of this was radiator hose. Typical Russian radiator hoses will survive for over thirty years without replacement; by 1992 hoses were bursting after six months because of a chronic shortfall in rubber supplies to the factory. Spare parts were so scarce that trade sprang up all over the world to ship old Lada cars back to Russia either as complete cars or as sources of spares. The Russian-registered car shown here was originally a British import, as the Lada Riva badge suggests. A trail of Lada shells is said to stretch from Hull to Murmansk, as stripped-out hulks were ditched overboard from trawlers and cargo ships. Other problems were associated with organised crime. A director of the factory was shot dead, shady middlemen wandered unchecked around the factory, and funds and spare parts haemorrhaged from its gates.

Problems such as these are now reportedly under control. News suggests that cashflow is back on track. Sales in Russia are so strong that at the time of writing (1999) all export contracts are indefinitely suspended to fulfil insatiable demand at home. The factory currently produces in excess of 800,000 cars, well up on previous post-Soviet years and approaching the figures of the record-breaking late '80s. The only question mark on the horizon for AvtoVAZ is its future as a partner of General Motors. This, I suspect, no one can predict at the present time.

As for the future, this must be bright for Russia's largest manufacturer. Who knows – we may even see a production sports car at some point, as a prototype was shown at the Moscow Motor Show in 2000 (below).

2

Ukraine, Estonia, Latvia

While the Ukraine is today an independent nation, back in the '60s it was an integral part of the USSR and was responsible for Russia's main economy car, the ZAZ. The LuAZ enjoys a rather unfortunate nickname in the former USSR – namely 'pansam sclepan', which means 'you build it yourself, sir'. The quality of vehicles coming out of the Lutsk plant was so abominable for many years that it was the only Soviet vehicle you could buy without having to sit on a waiting list. The model range was always very basic and generally used the standard ZAZ engine. The military 4×4 was designed to be air portable and was of ultra-simple construction, while the civilian version, the Volin, illustrated here, was designed for foresters, farmers and country people in general. Like all civilian vehicles, the Volin or LuAZ 969 was distributed via the Communist Party of the Soviet Union. When the showrooms received the latest batch of cars they divided them between large enterprises, local trade unions and a committee that compiled a list of workers who wanted to buy a car.

The Volin's abominable reputation meant that batches were often returned to the factory, and while workers would gladly have rebuilt them to a proper standard, the machines simply vanished. The largest quotas were allocated to the Ukraine, so Russians from as far afield as Siberia made the long trip to try to purchase a vehicle locally. The Volin is capable of carrying 880lb plus driver and a tow hitch is always included for a trailer. The machine weighs 1,980lb and the top speed is 56mph. With a compression ratio of 7.2:1 the poorest petrol can be consumed, and yet the economy is a very reasonable 30mpg. Excellent ground clearance (11in) is achieved by having the axles offset upwards, with a micro gearbox in each wheel and shaft drive from a central gearbox. Two-wheel drive can be selected for normal road use. Suspension is all independent with torsion bars, and thanks to a completely flat floor there is space for a lilo and a comfortable night's sleep. The ZAZ engine is situated under the bonnet, as is an effective petrol heater. The export version is driven by a Ford 1117cc engine with a 9.5:1 compression ratio. Power is 48hp compared to 34hp from the ZAZ engine. The Ford-powered version, launched in 1988, shows how it is possible to make a silk purse out of a sow's ear. Thanks to the diff lock and surprising ground clearance owing to the offset axles, this machine can go where the Niva (and many Western 4×4s) cannot. The current model is essentially the same as that which first rolled off the production line in 1972.

The military version (shown here) has the appearance of a motorised skateboard, particularly when the windscreen is folded flat: the vehicle is designed to be parachuted out of an Antonov. The air-cooled engine is excellent in low temperatures and is resistant to battle damage: there are no fluids to leak. A front winch, black-out headlamps and treadways are supplied as standard and the driving position is in the centre so that a stretcher can be carried either side. The steering column also collapses so that it is possible to drive while lying down. Why anyone should want to do this is unclear, although one Russian soldier mentioned that a prone driver between two stretcher cases was unlikely to be hit by any bullets. Either side of the driver there are two folding seats, for use if stretchers are not being carried. For military operations the LuAZ can carry an AGS-17 grenade launcher, the AT4 SPIGOT anti-tank guided missile and an 82mm B10 recoilless rifle. While the vehicle is fully amphibious, drivers should 'take care of the gaiters, ensuring that they are not torn', according to the handbook.

From the middle of the 1950s there was a realisation in the USSR that a people's car was needed, and in 1955 an engineer named Domatovski presented the Nami 012 (above), a rear-engined car of the most unusual design with an automatic gearbox, power-assisted steering and independent rear suspension. A year later the just as unusual Nami 50 'Bielka' with two side doors and a front door à la Isetta was shown to the public. Power was either a 500cc or 750cc flat twin in the rear.

In 1959 events took a step forward with the MZMA plant in Moscow demonstrating a people's car with a V4 air-cooled engine in the rear, torsion springing in the front and rear suspension of triangular units with coil springs: this had a technical similarity to a certain famous German machine. The company also created the Maljutka, shown here. This gorgeous bug-eyed monster dates from 1956 – but only a prototype was built.

In 1960 serial production began of the ZAZ 965 in the Kommunar Tractor Works on the river Dnepr. Tractor production was abandoned at the same time and the car was nicknamed Zaporozhets after the nearby town. The V4 air-cooled engine came from the Melitopol Engine Works, and is possibly the world's only V4 air-cooled car engine.

In 1961 the engine displacement was increased from 700cc to 900cc, and the new model was known as the ZAZ 965A. The styling similarity to the Fiat 600 can be seen in the recent James Bond film *Tomorrow Never Dies*, when Bond is picked up at Leningrad airport by a CIA agent in a Zapo, which then self-destructs.

In 1967 the ZAZ 966 was launched with a three-box body similar to the NSU Prinz. In 1968 the engine capacity was further expanded to 1200cc. The handling of the new model was improved considerably by a redesign of the rear suspension to parallel trailing arms rather than trailing arms hinged at the middle of the car.

A modest redesign occurred in 1972 and subsequent cars were known as ZAZ 968. Maximum capacity was 150,000 cars per year, and the production line was over 10 miles long.

To make the car more attractive in export markets, a 1000cc Renault engine could be ordered from an assembly line in Belgium. This model was known as the Yalta 1000. The ZAZ 968 had dual circuit brakes with discs at the front. A number of parts were shared with VAZ of Russia. Output peaked at 165,000 in 1977, of which 20 per cent were exported. The body shape remained broadly unchanged.

By 1983 more than two million Zapos had come off the production line, and the final version had a smoothed-out body without the rather attractive air scoops at the rear.

The Zapo driving experience is not altogether pleasurable. The V4 engine is a good tree stump puller but the top speed is low. Even Soviet advertising in the Western markets was modest enough to alert drivers to the fact that it could be driven at 'normal speed'. As one might expect, handling is similar to that of the Beetle and, also like the Beetle, the gearchange is surprisingly good considering the rod connection to the gearbox. The interior is spartan and the Zapo engineers never adopted the Beetle's heat exchanger, simply relying on a thirsty petrol heater. The front boot has a spare wheel, battery, petrol heater, steering box and steering column, and while the steering is vague, at least the column is collapsible in the event of an accident.

1986 saw the launch of the new Tavria model, the ZAZ 1102 – and this extraordinary picture is taken from the brochure. The first fifty front-wheel drive cars were shown at the twenty-seventh Communist Party conference, and most of the development work was conducted by VAZ at Togliatti. The 1100cc engine looks remarkably similar to the Lada Samara unit, but apparently very few parts are shared and the engine is once again made by the Melitopol works. The new car was more than 286lb lighter than the old and fuel consumption fell by over 30 per cent. Full production started in 1987, with the old and new models being manufactured side by side. In that year 968M production totalled 50,000, and 100,000 Tavrias were produced. Production soon rose to the old Zapo levels, but when the Ukraine became a separate state outside the CIS problems started to arise. Unlike VAZ, components were brought in from suppliers in Petropavlovsk, in Lutsk (where the four-wheel drive vehicles were built) and, of course, Melitopol where the engine is produced. Shortages of foreign currency meant that production remained at a standstill for long periods of time.

Lada UK planned to import large numbers of Tavrias in the early 1990s for sale at prices as low as £3,000. Because of production problems, however, nothing became of this and the three registered right-hand drive prototypes sat at the Lada Import Centre in Bridlington until 1999, when two were broken for scrap and the last was sold to the author for spares.

The late '90s also saw a liaison between ZAZ and Daewoo, with Daewoo aiming to invest $1.3 billion. The production target was over 250,000 vehicles, including 80,000 of the Tavria model per year. The Ukrainian government supported Daewoo by passing a law in March 1998 preventing the importation of vehicles over five years old. Despite that the marriage started to fall apart, and 1998 production totalled only 13,000 Daewoos assembled from kits and 11,000 Tavrias. Only 5,000 Tavrias and 6,000 Daewoos were sold: in 1960 the old Zapo had cost twenty times the average monthly pay, but by the '90s this had risen to eighty-seven times the typical monthly salary, so new cars just weren't affordable. Prices fell to below $1,000, making the Tavria the world's cheapest car for a while. Daewoo had problems of its own in Korea and the entire factory ground to a halt in December 1998. With a modest improvement in the Ukrainian economy in 1999, production restarted in March with Ukrainian companies supplying many of the parts formerly imported from Russia. Daewoo production did not restart because of the higher cost of these cars. The Tavria was re-launched as the ZAZ 1103 Slavuta, and the price tag for this five-door model is $3,500, $700 more than the standard Tavria.

With fewer than 1,000 cars produced in 1997 and 25,000 cars in 1998, 1999 production would appear to be around 500 cars per month.

The Tavria driving experience is not a bad one and you immediately notice the lightness of the car. For example, as with the Zapo, the wheelrims bolt on to the brake assembly saving the odd pound or two. The total weight of the car is just over 1,540lb and this results in excellent fuel economy of 62mpg. Front suspension is the McPherson type, with rear suspension being leaf springs. Maximum travel is an impressive 7in to take account of rough Ukrainian roads. There are one or two clever design features such as heat dissipation from the brakes through the wheel hubs, and a front seat which moves up when moved forward (as those who are short obviously need to sit higher). There is even a heater outlet on the floor of the rear passenger area. Weight saving has its penalty, with some of the fittings feeling shoddy – but the same can be said for Western European models costing twice the price. However, the styling of the van version leaves rather a lot to be desired.

Estonia is rarely mentioned in the motoring world, and yet over the last forty years it has been the mainstay of Soviet racing car production. Estonia also boasts a kit car manufacturer, Rex. The Rex R10 has a tubular chassis, fibreglass panels and is available in targa, coupé and cabrio form. Lada or Ford units can be fitted and the manufacturer is apparently a Mr Andres Roosma.

In 1999 Estonia boasted 458,000 cars on its roads, with 185,500 registered in Tallinn. Ladas are the most popular (109,000), followed by Ford (47,000) and Moskvich (44,000).

In 1957 a number of engineers at the Tallinn Auto Repairworks Number 1 decided to build a single-seater race car, and the Estonia 1 (above) was finished a year later. It set a lap record at a street race course in Leningrad.

The Estonia 1 would have appeared to have borrowed quite a number of parts from the Ural/Dnepr motorbike, which is, of course, BMW inspired. Even the wheels were pure bike with huge drum brakes fore and aft.

The Estonia 9 (above) was a far more sophisticated proposition, and thirty-nine of these machines were built in co-operation with Melkus, the racing division of Wartburg. The Estonia 15 came out in 1967, all 250cc worth. Production amounted to 221 cars. The Estonia 16 used the Moskvich 412 engine, while a heady 295 cars of the Estonia 21 model were built. Thanks to rally success and sponsorship by famous names such as Valvoline, the factory in Tallinn continues to prosper and welcomes visitors.

Latvia is probably better known for barn finds than vehicle production, for example a 1936 Auto Union with a 500hp engine, which is now in the West. Even so, Latvia has a long tradition of car production. Even before 1900 the Riga firm of Leitner was building three wheelers with 1.3bhp De Dion engines. The Russo Balt factory was busy producing lorries and cars before the First World War and a Russo Balt K fire engine is on display at the Riga Auto Museum. The Russo Balt seems to have died after the 1918 revolution.

Latvia's history of car production re-started on the back of a confidence trick. As the Second World War drew to an end, a car repair workshop which also made water heaters and orchard sprayers was nationalised. In 1947 a mysterious Russian named Vsevolod Bahchivandzhi arrived in Riga boasting that he was the brother of a famous Soviet test pilot. He claimed that the Soviet Airborne Division was interested in a prototype car he had designed. When the local authorities saw a chit signed by Comrade Stalin they capitulated, and the Riga Experimental Factory (REF) was constructed at the site of an old industrial plant on Alkasnu Street. REF was to be a broadly based concern manufacturing radio loudspeakers, metal furniture and the promised automobile. Time went by and no cars were produced. Eventually the local authorities smelt a rat, and decided that by the end of 1948 a prototype must be constructed. Proper funding was allocated with a recommendation that a Soviet engine be used. Milling machines, grinders and so forth were purchased, and yet after quite a number of months there was no sign of cars or loudspeakers, and the metal furniture turned out to be expensive and ugly. In the summer of 1949 Stalin signed an order that two automobile prototypes must be made by the anniversary of the October Revolution. Serious funding was provided – 600,000 roubles, and this in a period when the Moskvich 401 cost only 9,000 roubles. Riga civil servants were still highly nervous, especially when they found out that one of the REF founders, a Mr A.J. Kuzko, was a swindler who was wanted as far away as Manchuria. Work went ahead with much time being spent on automatic transmission. A home-made air-cooled engine failed, and then they turned to a GAZ engine – cutting the 4-cylinder block in half to form 2 cylinders. It was thought that Soviet paratroopers did not need any shock absorbers, so they weren't provided. Stalin's deadline grew ever closer, and no doubt many an engineer's buttocks were clenched at the thought of being shipped off to Siberia. Most surprisingly Moscow relented, and a year's grace was given. In the autumn of 1950 a convertible and a saloon were ready for testing at Moscow's auto institute laboratories.

The 620 mile trip from Riga to Moscow was in itself a serious enough test, and both cars made it. However, the Russians found one or two defects, for example an engine-out job involved removal of the entire body. The expensive automatic transmission worked imprecisely, and there were problems linking the radiator at the front with the engine in the rear. Finally, instead of about 40mpg, the fuel consumption was 23mpg. The Moskvich 401 appeared to the Russians to be a far better bet, so after more than one million roubles had been wasted they pulled the plug.

Bahchivandzhi's departure from Moscow was no doubt a gloomy affair, particularly when the convertible broke down and had to be abandoned along the route home. The saloon made it back to Riga and now rests at the Motor Museum there. Bahchivandzhi retired in Riga after various other hair-brained schemes such as micro air conditioner production failed to prosper.

That, however, is not the end of the story. Riga is far better known for its production of microbuses, and the most popular minibus in the Eastern bloc is the RAF Latvia. At the end of the Second World War Latvia was a devastated land, and out of the 160 buses that had been used before the war to transport the inhabitants of Riga, only ten were left. After the unsuccessful experiment with the paratrooper transport, the RAF factory started to produce trolley-buses, and then in 1953 the first new Riga model was produced – based on a GAZ 51 lorry. GAZ 51 trucks were received from Gorky and all superfluous parts were cut off, with a new body being grafted on. Tooling for the roof was not available, so this was canvas covered and the windows were thick windowpane glass.

In the mid-'50s RAF started to enjoy its renaissance, as the Soviets were producing nothing in size between the normal five-seater car and large buses. The Volkswagen seven-passenger microbus started things rolling in 1951, and in true Eastern bloc style, the Volkswagen microbus was brought to Riga and studied. The Riga engineers had no choice as to the engine: the only reasonably powerful Soviet car engine was that of the GAZ M20 Pobieda. A team of seven built the first microbus, engineers and designers mucking in with the workforce. The result was the handsome RAF 10 of 1957, pictured here, which resembles a scaled-up VW microbus even down to the split screen and roof windows. The main difference was of course that the engine was in the front. In 1958 the government placed an order with the factory to provide ten microbuses for the transportation needs of delegates attending the twenty-first congress of the Communist Party of the Soviet Union. Mass production was then ready to begin.

The closeness of the headlamps on the RAF prototype was abandoned in serial production.

In 1959 the model was redesigned with the more powerful GAZ 21 Volga engine (ohv rather than side valve). The look was still very Volkswagen. Work on the 'shovel front' model that shuttles Russian citizens around today began in 1963. Amazingly enough the grille and headlamps were simply taken from the Moskvich. Up to then, compared with Western microbuses the RAF machines were 30 per cent heavier, consumed more fuel and wore out more tyres. However, their working life on poor Soviet roads was longer. The new model attempted to overcome these defects and the Volga 24 engine was now available (in 95bhp form) to power it.

RAF were also involved in coachbuilding, and built the estate version of the Russian Chaika limousine. RAF started to die when the Baltic states gained their freedom. Russia then began to develop its own microbuses rather than relying on imports and it is probable that over the next ten years the famous shovel-front RAF Latvia will vanish from Russian roads. At the end of the 1980s money was poured into new projects, which developed into the 1991 RAF Roxanna, which was designed with help from IAD in the UK. RAF has negotiated with Western producers but is not prepared just to assemble. Sadly the factory's future must be in doubt.

3

Poland

Polish motoring history stretches back a long time, all the way to 1921 when the first Polski Fiats were produced. Poland's first home-grown machine was launched in 1923, known as the CWST1 (CWS standing for Central Warsaw Factory). This was a six-seater with a 3-litre engine producing a meagre 59bhp at 3000 revs. Eight hundred were produced, namely 500 saloons, 250 ambulances(!) and 50 police cars. Poland's first luxury car was produced at Panstwowe Zaklady. The luxe sport V8 boasted a 3880cc engine and 94bhp at 3600 revs. Production started in 1938 and soldiered on to 1941. The luxe sport was ahead of its time with good streamlining, and who knows what could have happened had the Germans not invaded.

In 1934 Polski Fiat launched the Fiat 508 series on the Polish market and very impressive they were too, with the 1 litre producing 18kw at 3600 revs. The Fiat 518 followed shortly after, with power upped to 44bhp at 3400 revs; 2,500 to 3,000 were produced per year. Within a short period of time the Fiat 500 Topolino (above) and Fiat 1100 and 1500cc models were launched, as was the Steyr 55 under licence from Austria. General Motors built a factory to produce Buick, in 1937, making a Buick 41 19 Master – as well as the Opel Kadett and Opel Olympia. Needless to say, operations ceased in 1939.

Like other Europeans the Poles experimented with microcars, and one of the more interesting machines was the Smyk, launched in 1957 by Wojciki at the Buirze Konstrukcyjnym Przemyslu Motorysacyjnego in Warsaw. Production was very limited with perhaps twenty machines being produced, although *The Motor* published a picture of the Smyk in 1957. The front door and steering column hinged forward in a most curious manner.

Rather more successful was the 300cc Meduza produced at Mielec by the Wytwornia Sprzetu Komunikacyjnego factory. The Meduza was featured for the first time in England in *The Motor*, August 1957. It was fitted with a two-stroke, 2-cylinder power unit at the rear with independent suspension on all four wheels. From the side there is a vague similarity to the Goggomobil. The same factory produced the Mikrus MR 300, with a rear engine and two-stroke power. An attractive cabrio version was made, although only in prototype form. There were quite a number of 'also rans' in the Polish microcar community, with the Gad 500 calling itself a sporting mini car. This was designed by Stefan Gajecki at the Warsztacie Doswiadczalnym Polskiego Zwiazku Motorwego, in Warsaw. The Gad 500 had its own 496cc two-stroke engine, which was also used to power the Pioneer micro-station wagon made at the workshops of the Ministry of Transport and Civil Aviation. The P80 three wheeler never made it to production, which is a shame given its charming looks. It was originally fitted with a WFM SO1 engine, but this was replaced by a Jawa 250cc unit. A WFM engine was also fitted to the FAFIK microcar built at Warsaw's motorcycle works. This too only existed in prototype form.

In motoring history one occasionally expects Western machinery to have an afterlife in the third world or Russia. But what happens when these nations decide that their cars need updating? The answer is, hand the old one on to a Soviet satellite. The M20 Warszawa started life as the Soviet Pobieda, and on 6 November 1951 the first car came off the production line, essentially screwed-together Soviet parts. 1951 production amounted to seventy-five cars; this rose to 1,600 in 1952. In 1953 engines were produced in Poland, and by 1956 the Warszawa was 100 per cent Polish. 1956 also marked a faint restyle, and the old side-valve engine was replaced by an ohc unit producing 70bhp at 3600rpm, giving a top speed of 65mph. Total weight was 2,992lb. In 1957 an estate (below) was produced in prototype form, while in 1958 a facelift took place: the main change was a less ornate grille.

In 1960 Ghia were asked to restyle the fastback rear, and came up with a rather ugly booted version.

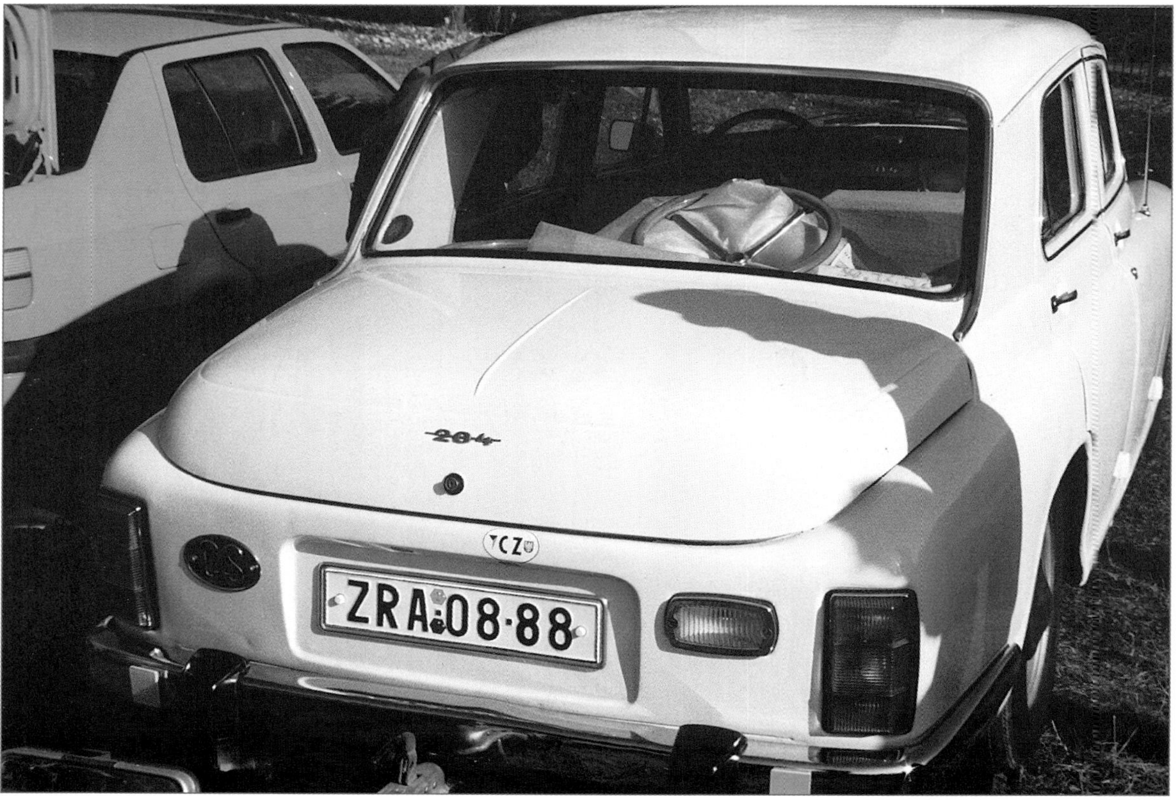

In the early '60s a van and ambulance were launched, annual production being still a meagre 20,000 units. Poland's Zuk Nysa and Tarpan Honker vans were based on the Warszawa chassis, while Warszawa body panels, lamps and indicators were used on the much smaller Syrena prototype. The Tarpan Honker is shown here.

Extraordinarily, the Pioneer microcar used one of the Warszawa's rear leaf springs, transverse in the front, as its suspension.

Surprisingly the M20 is monocoque with a front subframe for the engine's suspension. Until recently they were used as taxis and one taxi driver claims to have covered 1¼ million miles in his. Petrol rationing saved the M20, as the allocation for a large car such as this was more generous than for smaller machines. Thus many M20s were kept in running order simply to obtain a better ration. Exports were limited because of the car's aged design, although some went to China and Brazil.

The Warszawa 210 prototype was completed in 1964 and appears to have used Volga (2.5 litre) power. Sadly, funds were not available for its development.

On 30 March 1973 Warszawa production ceased, with a total of 254,470 vehicles having been made. The Poles are now starting to treasure the Warszawa. The M20 is an incredibly tough machine, ideally suited to Poland's abominable roads. Early versions were supplied with a tow hitch to enable an anti-tank gun to be towed on mobilisation. Independent front suspension with wishbones and a rigid rear axle hung from leaf springs is the system still used to this day, and the M20 has a number of surprising luxuries such as two-speed wipers, electric cigar lighter and adjustable radiator blinds operated from the driver's seat. A comprehensive toolkit was supplied, and mpg was around 22 in town and 28 on the open road.

The origins of Syrena date back to the 1940s, when the idea for a small car germinated within FSO's management. FSO's full name is never mentioned by non-Polish speakers, for obvious reasons (Fabryka Samochodow Osobowych, or passenger car factory). At the end of the Second World War Polski Fiat/FSO signed an agreement with Fiat, the plan being to produce Fiat 1100B and 1400 models for ten years from July 1948. It would appear, though, that the Soviets leaned on the Poles and forced them to tear up the treaty, and the proposed factory at Zeran started to produced the Warszawa. This left FSO with something of a problem, as the Warszawa was too large and upmarket to be suitable for middle management. The working classes, of course, were not expected to own cars. Plans started to develop in 1953 when the State Commission for Economic Planning asked FSO to develop a 750cc passenger car; 50 per cent of parts used were to come from the Warszawa. This was a tall order given the difference in size of the two cars. However, the first prototype was built in 1953 using a wooden frame covered with leatherette, a similar production method to that used in East Germany. Stanislaw Panczakiewicz designed the coachwork while Karol Pioneer designed the chassis. The 690cc engine was brought in from East Germany where it powered the IFA F8. Bhp was 22 at 3800rpm with a 6.1:1 compression. Maximum speed was 53mph. Prototype number 2 used parts from the Warszawa cut down to size, which astonishingly did not look out of place.

In 1954 the next design appeared, the Type 100. The Poles already had a two-stroke engine used for fire pumps and this unit was pressed into service: cubic capacity was 746cc and bhp rose to 27 with 6.8:1 compression ratio. Tested against a Panhard Dyna, a DKW and a Goliath, the car performed well, with a top speed of 64mph. At the 1955 Poznan fair a four-stroke prototype (model 107) was exhibited, with the engine based on the Junak motorcycle unit. Cubic capacity was 744cc with 35bhp available.

Full production began in 1957 with the model 101. The basic design was now frozen and changes over the next twenty years would be modest. Design features included a gear lever looking like a walking stick, also seen in DKW cars. The wooden frame soldiered on until 1962 when the model 102 was launched. Bhp crept up to 30 thanks to a higher compression ratio. Transverse leaf springs front and rear remained, however. The walking stick gear lever was replaced by a column change. Changes were also made to the grille. The 102S model boasted a 991cc Wartburg engine with 40bhp, and a top speed of 78mph. Model 103 followed soon after with barely any changes. Once again this model was also available with the Wartburg engine.

The 1960 Syrena Sport was displayed at the Brussels Autofair. It had a glassfibre body, and an air-cooled boxer engine.

The year 1960 saw the development of a seven-seater microbus of compact design similar to the West German Lloyd LT600.

The next model, 104, boasted a 3-cylinder rather than the original 2-cylinder engine. Cubic capacity was 842cc with bhp rising to 40 at 4300rpm. Compression was still a leisurely 7.2:1 because of low octane petrol. This engine was to last until the end of production.

This 104 was the highest-placed Polish entry in a 1960s Monte Carlo Rally, which isn't saying a lot.

In 1971 production of the Syrena was shifted to FSM in Bielsko Biala, presumably to release factory space for more modern machines. Commercial variants were introduced, such as the pickup and delivery van. The van could also be ordered as an estate car with windows all round.

The Syrena 105 was launched in 1972 and benefited from the end of suicide doors, a synchromesh gearbox and larger side windows.

The Syrena driving experience is an interesting one. The quality of the interior fittings is surprisingly good, although the steering wheel is huge. The window glass has the thickness of a shop window; however, large quarterlights provide efficient ventilation. The circular speedometer is quite Beetle-like, while the move from a column change to floor change was probably a mistake – although if progress is not hurried there are no problems. The large pedals are offset and sprout so far out of the floor that the driving position is not that comfortable. The engine is a little gem with a glorious turbine whine when under power. One odd little luxury is a switch enabling you to have either all or some or no instruments illuminated at night. As with other Polish cars the Syrena continues the tradition of hard springing/soft seats, ideal for bad roads. The seatbelt catch is very curious indeed and appears faintly aircraft inspired.

Because of the dated design, a number of reskin jobs were attempted, with the Syrena 110 being an example. Faintly reminiscent of the Renault 5, a two-stroke and a four-stroke prototype were built.

The Syrena 607 prototype benefited from a longer wheelbase and a rear hatch, together with plastic body parts.

Syrenas are now rare in Poland, particularly the saloon variant. Pick-ups are relatively popular and prices vary from around £50 to £200 for concours. While West German classic car enthusiasts are hoovering up anything old, the Syrena is not on their shopping list – and this poor little orphan is now mostly seen in scrapyards. Most Poles would rather forget the Syrena, but given that it has a chassis, two Warsaw students converted theirs into a cabrio.

Having given the Italians the push after the war, the Poles started to negotiate once again with Fiat in 1965, and in 1966 an agreement was signed to build the Fiat 125p for ten years. In November 1967 the first 125p left the assembly line, and whilst this was basically a car made from Italian parts, the percentage of the car made in Poland quickly rose, reaching 40 per cent in 1968 and 80 per cent in 1969. Quite a number of parts, including valves, electrics, instruments and brakes, came from Zastava in Yugoslavia. In the usual Soviet style of bartering, headlamps, rear lamps and electrics were sent to the Soviet Union in exchange for locks, bearings, windscreens and other parts.

One of the stranger versions produced was this four-wheel drive estate car. The axle stands appear to be holding it up.

The 125p was similar to its Italian parent only in shape. The Fiat 125 contained a new twin-cam 1600cc engine, which could propel the car to 100mph. It was essentially a Fiat 124 stretched fore and aft of the glasshouse. The Poles had to make do with the old Fiat 1300/1500 power train for the 125p, and the stylish chrome parts originally imported into Poland soon gave way to black plastic with the texture and smell of the typical bin liner. The 1481cc pushrod unit produces 70bhp at 5400rpm, while the carburettor is a Polish-made variable twin-choke down-draft Weber. The Italians ceased production of the Fiat 125 in 1972, by which time the Poles were getting into full swing. Their first major development was in 1971 with the launch of the 125p estate, shown here, a variant never available in Italy. In 1973 the pickup and van variants were launched. FSO also assembled other Fiat models, although never in great numbers (the Fiat 127, 128, 130, 131 and 132). Comecon economics in the '70s produced some weird variants such as the Z125p, made in Yugoslavia with a percentage of Polish parts. Another wasteful exercise was the assembly of Zastava 1100s in Poland.

What is the 125p like to drive? The answer is bad, but not that bad. For a start, the 125p (like its Russian counterpart the Lada) weighs rather more than the original Fiat version and the pushrod (rather than ohc) engine is a rough old beast. The Polish tradition of hard suspension and soft seating makes sense in a country with very rough roads, but less so in the UK. Likewise, poor fuel consumption and the fact that the engine is over the 1100cc barrier for cheap road tax means that a 125p with an MOT can be had for £100. At this price, the disadvantages start to fade away and, with quite a reasonable five-speed gearbox, it fulfils its purpose. Parts are stunningly cheap, and one oddity is the fact that when the front brake pads are part worn, they can be used to replace the rear pads. In 1977 Fiat successfully disassociated themselves from their Polish offspring by requiring them to drop the name Polski Fiat: this was replaced by FSO.

This six-door convertible version was probably unique.

Two prototypes from the Warsaw technical museum, the above built by FSO, while the machine below may have been a student project.

1978 marked the launch of the FSO Polonez. Basically a reskin, using the same floorpan, the Polly was 264lb heavier and had a price 20 per cent higher than the 125p. It was intended to make the Polly lighter than the 125p but this proved not to be possible. Although an attractive looker when it was launched, the Polly was remarkably labour intensive, requiring a large number of manual welds. The rear leaf springs were made slightly stiffer to cope with the extra weight, and small changes to the engine improved the bhp. Through the '80s a large number of detailed changes were made, the most significant being a revised front end and rear window styled by Ital. In 1987 the Polly 1600 was introduced, and a pickup version was also launched. Various different engines were tried, including a diesel based on the Polly block, and for a brief time VW diesel engines were available.

The Polly is surprisingly comfortable, with cloth seats, adjustable head restraints and Western luxuries such as four interior lights, door open warning reflectors, remote control tailgate, illuminated glovebox, clock, rev counter and so forth. The 125p continued in production, and indeed FSO had a good level of success in not just exporting but assembling cars in Egypt, Indonesia, Ireland, Yugoslavia, Columbia, Thailand and Malaysia! In the mid-'80s FSO started to develop a Polly replacement, and in October 1986 in Posnan showed off the front-wheel drive Wars, the intention being to manufacture this Fiat Ritmo (Strada) look-alike with 1100cc, 1300cc and 1600cc engines. FSO, being seriously short of cash, was unable to fund any further development. The last 125p left the factory on 26 June 1991, production totalling 1.2 million. The Polly died in 1999 because of falling sales and the introduction by Daewoo (FSO's current owner) of the Matiz. Even the introduction of a 1.9-litre Citroën diesel engine and Rover petrol engine could not hold off the Polly's demise.

The Polski Fiat 125 was launched in the UK in 1975. Sales in 1976 amounted to 656, well behind Lada at 1,762 but ahead of Moskvich at 206 cars. Priced at £1,449, the 125 was more expensive than Britain's cheapest car at the time, the Wartburg at £999, but compared favourably with the Vauxhall Viva at £1,615. The Polly was launched fairly shortly after its introduction in Poland and the first sales were in January 1979. In July 1981 the Polly received a five-speed gearbox, while the 1990 range included a 1.5 SCE (82bhp); the LSE and Prima had the 1.6 engine (87bhp). The importers, MCL Group Limited, pulled out in 1991 and FSO Cars UK relaunched the marque in May 1994. That year they sold 282 Caros (the new name), of which 220 were powered by the 1.9 diesel. The pickup sold slightly better, with 297 sales. Curiously, although the Rover-powered Caro was available in Europe, this version never made it into the UK. In 1997 Daewoo pulled the plug on exports, a year after buying FSO, afraid of damaging its own brand by associating it with the downmarket Polish marque. Sadly FSO UK stood a good chance of prospering given the Lada pull-out some months earlier, but they were not given a chance.

The booted Polly was renamed the Celina and sold modestly in Poland and the Netherlands. However, Daewoo were keen to kill off the whole range, and thus this is the least common of all variants.

FSM launched the Polski Fiat 126p in October 1971 from its factory in Biesko Biala. In 1974 production of the Fiat 127 also started, but ceased for some unknown reason after 3,000 units. By the mid-1970s, 25,000 126s were rolling off the production line each year. In 1976 the engine capacity was increased to 650cc and the wheelbase was lengthened by 4½in. 1977 saw larger bumpers and a fold-down rear seat, while in 1978 production peaked at 200,000 units. From 1979 onwards the Polish plant was the only producer. In 1982 a better exhaust enabled a 12 per cent saving in fuel consumption, and with 50 per cent of production exported the 126 was a useful foreign currency earner. An astonishing 1.7 million 126ps had been produced by 1986. In 1987 the Fiat 126BIS was launched with a 700cc water-cooled engine, rear tailgate and 13 inch wheels. This did not prevent the Poles experimenting with a number of odd variants, such as a front-wheel drive coupé, and a 650cc air-cooled diesel. The 126BIS died in 1992, which also marked the launch of the Cinquecento. Many Cinque drivers would be surprised to learn that their steed was manufactured exclusively in Poland. The machine above boasts fibreglass modifications to increase the loadspace.

The 126p van failed because of lack of load space over the rear engine.

The coupé 126p remained a one-off.

The 126p estate prototype never made production, again because of poor load space.

The owner of this home-built four door could not afford the paint to finish the job.

4

Czechoslovakia

A luxurious limo, larger than a Mercedes Benz S class with effortless V8 power and turning heads as standard? That will be the Tatra 613. If it had not been for the Tatra's inventor, the VW Beetle might never have been, and for those very few lucky people who own one, the engine note (Porsche 911 meets Detroit V8) is pure music. Not only has Tatra continually produced automobiles since 1897 (one of only six manufacturers to do so), it has pioneered swinging half axles, air-cooling and the backbone chassis. Hans Ledwinka was born in 1878. The Nesselsdorf Car Plant in Czech Koprivnice opened its doors in 1897, and by 1914 Hans was already designing four- and six-cylinder engines with hemi heads, slanted valves, ohc and four wheel braking, which was unusual at that time. The marque was renamed Tatra in 1919, after the mountain range. In the '20s Czechoslovakia was relatively prosperous. Looking at the statistics for 1929, the Praga was the most popular vehicle (8,454 vehicles) while Tatra came second with 5,034 cars; Skoda was third with 4,158 while Fiat was fourth with 2,371.

Tatra's first success story was the T11 light car. Ledwinka's 1921 design took into account the long frozen Czech winters and was powered by a 1036cc air-cooled overhead valve horizontally opposed twin producing a relaxed 13bhp. The engine was installed at the front of a central tube chassis, driving the rear axle through a conventional gearbox using swing axles that needed no universals. Crown wheels attached directly to the swing axles arched vertically about the centre of the drive pinion shaft. Simple, clever and efficient. Hans was also involved with other manufacturers and the Steyr 6-cylinder 3.3 litre 12/40 of 1920 was his design.

Following the T11, the T12 was put into production with a slightly more powerful engine, and 25,000 were built until the end of 1929.

Hans wanted to break into Germany but his 6-litre V12 T80 never made it in the German market and only twenty-five were built. The T80 had a number of advanced features for the time, such as 12-volt electrics and four gears.

Rather more successful was the T57, launched in 1930 with an 1150cc engine, four-speed box and weighing in at only 1,680lb. This was the first model imported into Britain by a Captain Fitzmaurice of St John's Wood, and *The Autocar* was impressed by the astounding ride quality: 'it is a form of springing which shows up to its best on the very worst kind of potholed byway that can be found . . . the car can be driven at 35mph to 40mph where with the vast majority of ordinary cars . . . one would slow to 20mph or at the most 25mph'. At £260 the T57 was simply too expensive. A one-off special was the Fitzmaurice Tatra, advertised at £595 and featuring 1484cc, a balanced crank, pillarless doors, adjustable rear seats, a petrol gauge, speedometer and electric wipers.

This is the interior of the T57 B. Note the fake animal-skin above the instruments.

Hans liked the idea of air-cooled engines but found that locating them at the front had a number of disadvantages, particularly that of noise. In 1931 he developed his first 2-cylinder rear engine air-cooled prototype, while in 1932 the V570 prototype was developed, with an 854cc engine (opposite, top). Describing the unique features of this machine, one is immediately drawn to a rather more famous vehicle built for the masses of Germany: swing axle, central tube frame, torsion bars, a horizontally opposed air-cooled engine, and streamlining. The benefits of siting the engine in the rear were not just sound deadening but better traction in the snow, and simple removal of the engine. Air cooling also had benefits, particularly in winter with no fluid to freeze, and in summer when the cooling fan increased in speed in line with higher revs. With no hoses or pumps, the engine also proved to be lighter. Hans met Ferry Porsche a number of times in the 1930s and a Porsche prototype produced then had many similarities to the Tatra V570. Hitler admired Tatra and was driven in a T11 during his political campaigning. Hans's son Eric claimed that the pair formulated the basic design of the VW at a late night session; could Hitler have passed his ideas to Porsche? Tatra took out several claims against VW for patent infringement including ducted air-cooling, engine position and gearbox layout. Eventually VW came to a large out-of-court settlement in the mid-1960s. Sadly Hans never benefited from this, and he died in 1967.

The T77 (opposite, bottom) launched in 1934 was definitely not a people's car. A new engine was developed, namely a V8 of 77mm × 84mm bore and stroke, 2970cc with push-rod overhead valves, a 5.9:1 compression ratio and 60bhp at 3500rpm. Drivers either sat in the centre or on the right, given that right-hand drive was the norm until the Nazis invaded. The front screen was angled at 45 degrees to achieve wind deflection. There were no running boards and hinges were set into the car body. Maximum speed was 90mph. The carburettor could have cold external air or warm internal air. In 1935 a larger 3.4 litre engine was introduced, and the stylish lines made the Chrysler Airflow look like an old barge in comparison. Captain Fitzmaurice advertised the T77 in the UK for £990 but few were sold, as it was still too pricey. One T77 was crashed and some enterprising British enthusiasts built a mid-engined sports car using the engine and gearbox. Test day arrived, and much to the onlookers' chagrin the car hurtled backwards, having four reverse gears and one forward. Hans was less successful with the T77's weight, and with close to 2 tons to haul acceleration was dull. Other problems included front-end lift at high speed and catastrophic oversteer.

The T87 solved a number of the T77 problems. To cut the high polar movement of inertia, there was a new light alloy V8 which made the car 24 per cent lighter and also quieter. Rear visibility remained poor but in general lines were more graceful, with a flat windscreen with small panels each side. Perhaps Hans would have preferred a wrap-around had the glass technology existed. Compression ratio was only 5.6:1 and fuel consumption was a reasonable 22mpg. Top speed was 100mph from 75bhp. The T87 could happily cruise at 85mph.

The T87 was an effective luxury five-seater cruiser, but in 1937 Tatra launched the T97, which was far cheaper to build and run. Dimensions were more compact and the power plant was an air-cooled flat four of 75mm × 99mm, ie, 1750cc and 40bhp. Of simple unitary construction, it would have been a world-beater, but the Nazis stopped production because of similarities with the VW KDF wagen. Tatra exhibited the T87 at the Berlin Auto Show of 1939 but were told not to display the T57 and T97. During the Second World War trucks were produced in great numbers, although a number of T87s were built and supplied to Wehrmacht officers. Often, when taking their T87s out for a quick spin, they did just that, and shortly afterwards the German Army was banned from using the T87. The Germans changed Czechoslovakia to left-hand drive and this was reflected in Tatra production. Incidentally, Vauxhall examined a T87 in 1946 and were impressed. In 1979 a T87 was tested at the VW wind tunnel and achieved a cd of 0.36, a pretty impressive figure.

The new Communist authorities nationalised Tatra in 1946 and imprisoned Hans for crimes against the people. Production resumed with a T87 and an updated T57, although in 1948 the Tatraplan T600 appeared (pictured in an advert, and below with a T603 on the left). The T97 had its weight reduced to 2,640lb and a new 1950cc engine produced 52bhp. Torsion bar springing at the rear was introduced, although the swing axles were retained, and the top three gears were synchromeshed with a column change. 45 per cent of the weight was at the front. Ledwinka was released from prison in 1951 and left for Austria with his family. His son Fritz became a sales director at Steyr, while Eric went on to design the Haflinger and the Pinzgauer. What was the logic in changing the numbering system, in other words going straight from 97 to T600? As Tatra made everything from buses to rail vehicles and heavy machinery, numbers beyond 600 were allocated to cars while numbers beyond 500 were allocated to buses, and trolleybuses were numbered from 400. Even rail vehicles had their own numbering system, running from 300, while vans took 200 and small trucks 100. A number of interesting prototypes were produced, such as a two-door aluminium cabrio built by Sodomka in 1949. This was allocated the number 601. The Tatraplan T600 continued in production until 1954, although trucks were Tatra's main *raison d'être* at this time. No cars were produced in 1955 or '56, while in 1957 the T603 was launched. In case you were wondering what became of 602, this sports prototype was built in 1949 with an 83bhp engine.

The T603 is Tatra's best-known product with its flying saucer styling. The first prototype appeared in 1955 with a 2.5-litre 94bhp engine. Bhp rose to 98 and then 105, but an engine in 150bhp form powered the rally car that took third and fourth places in the classic Marathon de la Route in the early '60s. The T603 styling is unique with bulbous streamlining, a long and rounded nose and sweeping fastback tail. The massive wrap-around rear screen is split and great scoops bring air to the blowers that cool the engine. To cope with the weight imbalance, suspension is all independent with long trailing arms at the rear and coil spring damper struts, while at the front short trailing links located the wheels with springing à la MacPherson. The interior is huge with a domed roof giving ample headroom, and limousine-style legroom in the back. The floor is completely flat, enabling three people to sit on the front bench seat thanks to a dash-mounted umbrella handle handbrake and column change. Instrumentation is simple but clear. After priming the twin carbs, the V8 engine bellows into action but soon settles down to a quieter idle. The column change takes getting used to, with Tatra placing first and reverse away from the driver instead of towards him. The 603 is quiet on the cruise with up to 30mpg achievable on a long run. Steering is direct and light while handling is, to say the least, interesting.

The T603 above was photographed in the Yorkshire Dales, while the 603 below (with personalised numberplate) has the early close-set quad headlamps.

T603 production ended in 1975 after a run of 20,422 vehicles, and shortly afterwards the T613 was introduced.

The T613 was styled by Vignale, and with the 603 engine expanded to 3.5 litres, bhp rose to 160 at 5200rpm. This is slightly higher than the Rover 3500. The T613, with a long flat nose and short tail, is impressively styled although the near vertical sides make it look a little heavy. Weight distribution is improved as the final drive is installed beneath the forward part of the sump, the gearbox being mounted forward of the rear wheels. The twin ohc is belt driven with one camshaft for each of the two banks of cylinders. Suspension was largely carried over from the 603 and ventilated discs are fitted all round. The tyres are generally Barum 215/70 hr-14. For some odd reason the car uses two 6-volt batteries rather than one 12-volt, about the only similarity to Britain's MGB. The T613 is a very pleasant drive with a light clutch and a nice action gear lever; comprehensive instrumentation includes rev counter and oil temperature gauges. The red line starts at 5500rpm. 0–60 takes 12½ seconds, while a 100mph cruise is comfortable and the top speed is quoted as 118mph. Handling is good; however, small changes of direction result in the car weaving slightly as you come back to the straight-ahead position. The steering is not marvellous, with a dead feel. While the ride is very comfortable, the trim is fairly spartan and fuel economy is perhaps 14–15mpg in towns and 17–20mpg on the motorway.

Demand for the 613 was always low, with production slipping below two figures in 1994. Sales to Eastern Bloc countries simply evaporated, and after the Velvet Revolution the Czech government did not want to be seen squandering taxpayers' money. The new T700 was built strictly to order and the first model came off the production line in April 1996. Bhp rose to 197 at 5750rpm while the 0–62mph figure fell to 10.8 seconds. Weight distribution is 45/55 front to rear and power rack and pinion steering is provided, although it is rather over sensitive. The engine sounds magnificent, however, and if the T700 didn't weigh 4,056lb it would feel quick. Undoubtedly the T700 is the best air-cooled V8 rear engined car in the world (and also, of course, the worst). Tatra had intended to develop an entirely new car in 1998 but a lack of funds prevented this, and production of the T700 declined from twenty-three in 1997 to four in 1998. Officially T700 production ceased in July 1998, the factory losing money on every model: the price tag was fixed at $55,000.

Tim Bishop, a British engineer, attempted to import Tatras into the UK. He developed a fuel injection system and a host of other improvements. Maybe three or four British specification right-hand drive Tatra 613s were built with Connolly leather, air conditioning, and a number of unusual features such as a weird lightbulb check, which involved the on-board computer lighting up every bulb in the vehicle one after the other in an unusual light display.

A Tatra car is still in production, namely the Beta. This owes nothing to the legacy of Hans Ledwinka but is simply a small van built in conjunction with Skoda. Power is a 1341cc Hyundai engine with 84bhp. The van is selling in limited numbers in the Czech Republic and can carry 600kg. The body is GRP (coloured in red, white or blue). Designed in Ejpovice in the Czech Republic by Skoda designers, the manufacturing plant is in Pribor, with Skoda owning 60 per cent and Tatra owning 40 per cent of the shares. Pribor is the Tatra plant used to produce the T700 until 1997, when 'production' moved to Koprivnice.

The Czech government set the microcar ball rolling in 1948 when the Central Technical Commission announced a competition to design the best people's car. Requirements included two seater minimum, and maximum fuel consumption of 56mpg. A total of twenty-two prototypes were entered and most used motorcycle mechanics, namely Jawa and CZ. The smallest engine was 97cc and the largest 600cc. Most vehicles were open with a soft-top to keep down weight. As well as four wheelers, there were three wheelers and one weatherproof motorcycle (made by apprentices of the Karosa factory in Vysoke Myto, Eastern Bohemia). The microcars were exhibited in Prague but the Commission closed the competition without a winner.

The Czechs had not long to wait before the founding of Velorex, which was such a runaway success that there were virtually no attempts at competition. The Czech Velorex has been called many things, although luxurious is not a word that immediately springs to mind. In the early years of production it was nicknamed Hadraplan. At the time the most up-market vehicle you could buy was the Tatraplan, and the Czech word for fabric is Hadra.

After the Second World War the Czechs, like the Hungarians, decided to improvise given the shortage of vehicles, and quite a number of one-offs were made in private garages. These often consisted of a steel framework around a motorcycle, with canvas stretched over the frame for weatherproofing. Some machines were designed for invalids from the Second World War, while some of the brave drivers who drove these motorised wet fag packets became invalids for a different reason. One of these small garages was based in a village called Parnik. The owner, Frantisek Stransky, and his brother Mojmir built a number of one-offs with power ranging from a Manet 98cc to a Pal 300cc engine. Their next attempt was the basis of the Velorex, 12ft 11in long with a weight of 594lb. Power came from a 175cc CZ engine which enabled the trike to go up to a hair-raising 37mph. A dynastart enabled the two-stroke engine to move in reverse, meaning that there were four rear speeds as well as four forward speeds. Six-volt electrics and 19in wheels came as standard, and the leather cloth 'galon' stretched over the tubular frame was reminiscent of the motorcycle sidecars of the period.

In 1950 the government took over the factory and full-scale production was started in Hradac Kralove by the Velo Co-Operative. Early machines had a 250cc engine with 16in wheels, and production figures for 1951 were around 120, while in 1952 the figure rose to 180. This machine was known as the Oskar. Soon a 175cc engine was introduced. In 1954 the Oskar became known as the Velorex Oskar, and by 1959 production was close to 120 vehicles per month. Disabled people were allowed to buy the vehicle at a discount; the engines were properly run in before delivery, and hand controls were supplied. History does not explain how the disabled person was supposed to get into the vehicle, given the narrow door aperture. The Velorex Oskar even started to participate in track events, and the production chief engineer Jaroslav Ehl had a number of successes.

In 1953 development started on a prototype with a larger engine, a Jawa 350cc. Sadly, on 25 January 1954 Frantisek discovered the risk of driving a motorised canoe when he hit a tree and was killed. It was several more years before the 350cc was introduced. 1963 saw the full-scale introduction of the 350cc engine; the 250cc version was deleted. The 350cc version had a number of improvements such as a wooden floor and bench seat, with insulation to contain the noise and the heat of the engine. Like the previous version, tubular framed doors were supplied, as was a canvas roof and removable side screens. The body was suspended at the front on an axle using Jawa motorcycle wheels and motorcycle shock absorbers. The steering was by rack and pinion and the mudguards were also lifted off a Jawa motorcycle. Braking was supplied to all three wheels and was foot operated. The parking brake operated on the front wheels. Small ducts at the rear cool the rear engine, although the main cooling is provided by a fan run off the crankshaft. The standard Jawa 350cc engine was re-engineered by Velorex who produced an extended crank to run the pulley for air cooling. The fuel tank is simply a motorcycle unit welded to the sub-frame at the front.

Four-wheel prototypes were produced in the late '50s but these did not go into production. In 1960 they tried again and ten machines were produced. However, the four-wheeled version only came into its own in 1971 when three-wheel production ceased. This machine was called the Velorex 435, and 1,380 machines were made before production finally ceased in 1974. Around half the Velorex production was exported to Poland, while other machines were exported to Hungary and Bulgaria. Sadly all the machinery was broken up, although the name of Velorex continues as the company now makes fibreglass sidecars and car trailers. Production figures were approximately 800 of the 175cc version, 2,500 of the 250cc version and 12,000 of the 350cc version (excluding four wheelers). Today there are perhaps 1,000 left in the Czech Republic and Slovakia, of which perhaps 250 are roadworthy.

This customised Velorex has, no doubt, a modern Japanese motorbike engine at the rear.

Skoda has probably the brightest future of Eastern Bloc car manufacturers. While the future is bright it is also somewhat depressing, as today's Skoda is pure VW; even the free revving wet-liner Skoda engine has been confined to the dustbin. Today's Skodas are undoubtedly fine cars, but their heritage is German not Czech.

Skoda's history stretches back to the nineteenth century when two engineers named Klement and Laurin got together in 1894 to produce bicycles. Klement was aged 36 while Laurin was 29. In 1898 they produced their first motorcycle and full-scale production started in 1899 in Mlada Boleslav, 35 miles north of Prague. The first significant car was their Type A, shown here, which lasted from 1905 to 1911. It was a 2-cylinder 7bhp two seater using an 1100cc V twin. There were a number of rally successes, in particular the 1908 St Petersburg to Moscow event. Laurin and Klement experimented with some odd engines: how about a bore of 85mm and a stroke of 250mm? An agency was set up in the UK in 1907 at 255 Tottenham Court Road, and in 1909 the 10hp model sold for £195. Skodovy Zavody Plzen (Skoda Works Pilsen) took over Laurin and Klement in the 1920s and added car manufacturing to its wide range of businesses. In the early 1930s Skoda promoted its cars with long distance journeys, and one travelled around the world in a very respectable ninety-seven days. A Skoda Popular journeyed from Prague to India.

Skoda's engineering prowess ranged from heavy guns to locomotives and even 450bhp aero engines. They even produced Hispano Suizas under licence. Like a number of other manufacturers in the 1930s, Skoda developed a people's car prototype 932 in 1932. An air-cooled flat four of 1498cc was positioned in the rear. Strangely familiar? In 1935 a streamliner was produced, while even an incredible streamlined bus with every luxury was designed by Sodomka in 1938. Sodomka also built stunning specials for Skoda, and indeed other marques such as Aero and even Rolls-Royce and LaSalle. Skoda was well and truly a large mass-producer of cars, with names like Popular, Rapid, Favorit and Tudor in the 1930s.

On 1 October 1938 Hitler occupied the Sudetenland and in 1941 Skoda was incorporated into Goering's 'Reichwerke'. Pre-war Populars were still produced until 1946, while 1940 saw the launch of an ohv engine designed to replace the original side valve Popular engine. The fundamentals of this 1809cc engine were to remain in production for a further fifty years. A few Superbs with a 3990cc 96bhp engine were constructed, but production concentrated on the German war machine – with Skoda supplying parts for Tiger tanks and even producing light tanks of its own design. The war only really touched Skoda on 25 April 1945, when the Luftwaffe bombed the Plzen complex to destruction, seemingly simply out of spite. This was a sad day as the Skoda archives and a number of historic machines were also lost.

After the war the Skoda car operation was renamed AZNP, i.e. Automobile Works National Enterprise, and the Mlada Boleslav works was turned exclusively to car production while truck production was transferred to Skoda Plzen. AZNP was a nationalised firm, and the Skoda marque continued. First off the production line in autumn 1945 was the pre-war Popular model, pictured here and overleaf, with a four-speed box and the new 1101 ohv engine. The model was named the Skoda 1101 to underline the difference from the original 420 Popular of 1933 on which it was based. The 1101 came as either a two door or a four door or van/ambulance. Very soon the model was nicknamed the Tudor (two door) and the 1089cc ohv engine produced 32bhp, the bore being 68mm and the stroke 75mm. Unusually the gearbox is part of the engine block. The wheelbase and track were enlarged on the pre-war model and a new grille was pleasing to the eye. It was also pleasing to Renault who copied it on one of their later models. A four-wheel drive version of the 1101 was produced with a Jeep-type body, while the only other machine produced was the Superb, in very limited production for Communist party officials.

The Czechs very successfully built the business up and by 1947 exports had started to the rest of Europe, Africa and Australia. Czechoslovakia briefly benefited from the Marshall Aid plan, but in February 1948 the Communists took over the still democratic parliament and aid dried up.

The Czechs were also pretty quick in resuming their racing programme, and a Skoda Sport 2 racer took part in the Le Mans twenty-four race in 1951, with power boosted by a Rootes R015 compressor. Two cigar-shaped racers were also produced, known as the Supersport. They had a V-shaped cylinder head and 120bhp on tap at 6000rpm. In 1952 a larger engine was inserted into these machines, with 1490cc and two compressors producing an amazing 150 to 180 bhp.

Back on the production line improvements came through more slowly, with 1949 seeing the gearchange shifted to the column and estate and roadster versions offered. 1952 saw a genuinely new model (above) with a larger engine, bore being enlarged to 72mm and stroke remaining the same at 75mm; this 1221cc engine produced 36bhp. The public soon nicknamed the model the Sedan.

The Skoda 1200 was displayed at the London Motor Show of 1954, and Skodas were imported into Britain from that date – with Motokov taking over from the British concessionaires in 1958.

One problem with the Sedan was that it was certainly not a people's car, so in 1953 a prototype Spartak was produced with 995cc and 30.5bhp. In 1954 full production started and the model was known as the 440 – namely 4 cylinders and 40bhp. Another development that year was the introduction of the downdraft carburettor on the 1089cc engine. From 1995 onwards the 440 model was produced in great numbers, while the 1200 was given a 45bhp engine and remained the 1201. A 'hot' 440 had the 1201 engine and was known as the 445. Even hotter was the introduction of twin carburettors producing an extra 5bhp, and this engine was used in the 450 cabrio.

Racing successes continued, and in 1958 a racing roadster was built with a twin ohc engine developing 92bhp at 7300rpm. The 1089cc engine had a 9.5:1 compression ratio with hemi combustion chambers and inclined valves. Two twin-choke 35mm Jikov carburettors were used. To ensure reliability, there were two spark plugs per cylinder with a magneto each. There was even a five-speed box. The chassis was of tubular frame with GRP panels being used to keep the weight around 1,320lb. Maximum speed was an astonishing 155mph. One of the two machines built was driven to the UK in 1969 and bought by a British classic car enthusiast who restored it. Skoda eventually bought back the car pictured above in 1999.

Developments at the factory continued, and in 1959 the transverse leaf suspension was replaced by coils. The 440 model with new suspension was named the Octavia (above is the convertible) as it was eighth in the line-up of the 1-litre model range (Popular 420, Popular 995, Popular 1100ohv, Tudor 1101, Sedan 1200, Spartak 440, Octavia), the management having accepted the public's wish for names rather than numbers. They named the 445 the Octavia Super, and the 450 became the Felicia. In 1961 the Octavia was sold at £589 in Britain, while the Felicia was sold for £785. An Octavia TS was soon launched, namely an Octavia with a 50bhp Felicia engine, higher lift cam, new manifolds, bigger valves and higher compression. This engine in the Felicia made it the Felicia Super. Rally successes continued, and Skoda won first and second place in the 1001 to 1300cc class in the 1961 Monte Carlo rally. Skoda's first car was sixth overall out of 305 starters. Production could barely keep up with demand, and in 1960 a colossal new complex was built with a rail link; the main aim was to reduce dependency on outside suppliers, which had caused problems in the past. The new re-engined models were on the drawing board, and it was realised that volumes would have to be increased dramatically when these reached production.

Skoda realised that little had changed from 1933 to 1964, apart from the body styling, and from a blank sheet of paper designers settled on a rear-engined machine. This was because of the ease of building a monocoque chassis, and the advantages of a flat floor and the lack of propshaft. A water-cooled engine was preferred over an air-cooled engine because of lower noise levels and better heating in cold central European winters. The rest of Europe also chose the air-cooled route: VW with the Beetle, Renault with the 4CV and Fiat with its 500/600 range. The first 1000MB left the factory in 1964, with the 998cc engine inclined at 30 degrees to lower the height. The aluminium engine was mounted behind the final drive and boasted wet liners and a cast-iron head. While MB stood for the initials of the factory, the 1000MB was often known as '1000 small ailments', presumably because of quality control problems – or because the Czech words meaning 'small ailments' start with an M and a B. This new engine had a bore and stroke of 68mm and the power was 48bhp at 4750 revs. Compression was only 8.3:1, as the best fuel available was 84 octane. The radiator had a nine-blade fan and was placed on the side of the car, while the rear wheels were on trailing arms with coil springs and the front suspension included an anti-roll bar. The body tub was very strong and fuel economy was 40mpg, with a maximum speed of 78mph.

In 1966 the two-door 1000MBX variant was produced, with twin carburettors and a pillarless construction (opposite, below). The 1000MBG variant was the standard saloon with twin carburettors from the MBX. 1967 saw the engine increased to 1107cc by increasing the bore from 68 to 72mm. 54bhp was produced. One final variant was the 1000MB Deluxe, with an 8.5:1 compression ratio and a maximum speed of 84mph. The MB range was rallied successfully, with 1000MBs coming first and third in class in the Australian Alps rally. In the mid-'60s a 'hot' 1000MB of 998cc produced an astonishing 90bhp at 8600rpm using a Nitrided crank, forged conrods, Weber carb and oil cooler. Wishbone suspension and a cigar body completed the picture. This 130mph masterpiece is now in the National Technical Museum in Prague. The 1969 RAC rally saw Skoda first in the group 2 class (up to 1150cc) and for the next twenty years Skoda participated, winning in their class eighteen times. The MB bowed out in 1969, after over half a million had been produced.

Skoda felt a new body was needed for the MB, but the engine was carried over. The new model had front discs and twin circuit hydraulic brakes. Maximum speed was still 78mph, but the interior was improved. The 110L had a 8.8:1 compression ratio and produced 53bhp. This was the time when Skoda's famous coupé, the S110R (above), was launched. It was built not in the main factory but at Kvasiny. Between 1970 and 1981, 56,000 coupés were produced and 62bhp was extracted from the 1107cc engine. The coupé engine could also be specified in the saloon, the model being known as the 110LS. Rally developments continued with the 110L Rallye, the 1107cc engine producing 73bhp (or 100bhp for the factory team). The 120S 1173cc engine produced 64bhp, but once again the factory team extracted 120bhp at 7500rpm. Production increased to such an extent that 1969 saw the one-millionth car and 1973 the next millionth. By 1975, 650 cars were coming off the production line per day and 70 per cent were exported. Like Subaru, the rally successes must have helped sales. The 1975 rally season saw the 130 RS Rallye with an alloy head, twin Webers and 1289cc. The compression ratio was 10.5:1, with 112bhp at 7250rpm. Later, 129bhp was extracted. Body panels were a mixture of aluminium and GRP, and the maximum speed was 105mph – although fuel was guzzled at 19mpg.

The Estelle range was launched in 1976, with the 105 having the 100 engine and the 120 the 110 engine. The petrol tank was now over the rear axle with the radiator at the front of the car. The 105 engine was boosted to 1046cc, from a 68/72mm bore and stroke, with 46bhp and a maximum speed of 81mph. The 120 engine produced 52 and later 58bhp. In Britain in 1983 the 105S cost £2,378. Models available were the 120L, 120LE, 120LS, 120LSE and the Rapid, most expensive at £3,388. The Skoda Rapid was the replacement for the S110R, and was only known as the Rapid in export markets. Still built outside the main works, the Rapid had the new rear independent suspension, which eliminated criticism of Skoda's killer oversteer from journalists around the world. The Rapid's 1173cc engine produced 58bhp at 5200rpm. A 9.5:1 compression ratio and rack and pinion steering completed the picture.

There were further changes in 1984, with the new Rapid being launched with a 1289cc engine, bore being 75.5mm and stroke 72mm. 58bhp was extracted and a five-speed gearbox was standard. The 120LX was essentially a 120LS with new suspension and five-speed box. The 130L saloon had the 1289cc engine, and once again the new suspension. The British press gave the new product the credit it deserved, one headline being 'What handles like a Porsche 911, costs only £4,200 and is more fun than a GTi?' The very last new rear-engined model was the 136 Rapid. The 1289cc engine was continued but with a high compression 8 port alloy head. The inlet manifold was water heated for better starting. *Autocar* waxed lyrical: 'it offers a genuinely different driving experience'; and sales to Britain rose to a heady 16,000 in 1988. Lombard successes continued, and no doubt the noise from a Skoda at 8000rpm with a straight-through exhaust caused ripples of anticipation among onlookers.

The rear-engined cars had now been in production for nineteen years and change was needed. Skoda decided to keep the willing and eager wet liner engine but with help from Bertone on the styling, Porsche on the suspension and Lucas Girling on the brakes, the Favorit was launched in 1989. To enable 800 cars per day to be produced 102 robots were installed. The Favorit 136 came with the 9.7:1 compression ratio, Pierburg carburettor, 61.7bhp at 5000rpm and 93mph top speed. The range had four levels of trim, namely Forum, 136L, 136LX and 136LS. 1990 saw the estate version, the Forum LS and pickup, while a low compression engine was also introduced in the Favorit 135LS. Reviews in Britain were very complimentary, with the *Autocar* review for November 1990 complimenting the equipment, handling and the ride. *What Car?* named the Favorit the best budget car of 1991. In 1990 Skoda just missed the Lombard RAC class trophy, but won it in 1991 with a six-speed Favorit that produced 100bhp. The car's weight was only 1,747lb.

At the end of 1989 the Velvet Revolution swept the Communists from power, and AZNP was privatised – soon to marry VW in December 1990. The very last rear-engined car was produced on 28 December 1990. At about this time Czech engineers designed a new ohc engine of 1596cc with a five-bearing crank and multipoint fuel injection. The engine was ready in 1991, but VW engineers were not interested. The VW marriage had mixed results: although there was a colossal capital investment, there was also a severe loss of jobs. With the death of the Rapid, the branch factory started to produce the Skoda Forman estate and the pickup. Single point fuel injection and catalyst were introduced in 1993, while on 16 June 1994 the millionth Favorit had been produced (76,200 saloons, 190,000 estates and 48,000 pickups).

After three years' development the Felicia was launched in October 1994. The good old 1289cc engine survived in 54bhp and 67bhp form while a 1.6 litre VW engine producing 74 was available, as was the 1.9 diesel with 63bhp. Amazingly, the 1289cc engine was still used in rallying and even beat 2-litre machinery. For the purist, Skoda died with the death of the 1289cc engine in the late '90s, as current production is purely VW powered.

Avia, 1956. This 345cc Czech microcar did not know whether it was coming or going – and was a one-off.

The Lidovka was built in 1946, and is still in existence. It was powered by a 500cc DKW engine and had a 56mph top speed. Fuel consumption was 60mpg.

5

East Germany

Why has the Trabant become one of the most loved and loathed of Eastern Bloc cars? Firstly it is remarkably cheap. In the early days after the Berlin Wall fell in 1989, Trabis were being abandoned by their East German owners who were anxious to lay their hands on second- or third-hand BMWs. Ten years on the mint Trabis of 1989 are also being abandoned, having passed through the hands of innumerable German students. Today in Germany the Trabi is very much a poor man's VW Beetle, affordable transport with a certain amount of character. In Britain it is the bane of the DVLA who seem to make it as difficult as possible for Brits to import. An unregistered Trabi can be had for £150 in Britain, while a mint version should cost no more than £900. In Germany a mint Trabi would cost perhaps £350. Apart from being very cheap, Trabis are good daily transport for short distances. Reliability comes from a lack of moving parts (there are only nine), and, with most late model Trabis boasting electronic ignition, there is virtually nothing to go wrong.

The history of the Trabi goes back to the nineteenth century, when in October 1899 a German named Augustus Horch founded a small car plant in Cologne. In 1904 the factory moved to Zwickau where the conditions were ideal; the town's traditional textile industry employed many on low wages and this meant a pool of willing labour. Horch was dismissed from his own business in June 1909 for lacking the business skills to bring his visions to fruition. Horch refused to take his sacking lying down and within a month founded Audi Automobilwerke. The name Audi resulted from the Latinised imperative of the German verb *horchen*. Through the '20s car production continued, with both factories starting to learn the conveyor belt ideas introduced by Henry Ford in the USA. In 1929 a Dane acquired a majority share holding in Audi. He had already been successful with DKW motorbikes (Das Kleine Wunder). Using a DKW two-stroke engine he developed the F1, which was launched at the Berlin Motorshow of 1931. The F1 could probably be called the Trabant's oldest relative, being small, with a two-stroke engine and front-wheel drive.

Germany entered a period of economic collapse in the 1930s and various car makers were forced to amalgamate, given declines in sales. Auto Union comprised Horch (high performance prestige cars), Audi (front-wheel drive cars), Wanderer (middle range) and DKW (motorbikes and economy cars). Horch's ultimate machine had 12 cylinders, was 17ft 8in long and weighed 2.3 tons. This developed 120bhp from 6000cc. The top speed was 87mph, with a horrendous fuel consumption of 11mpg. Europe's cheapest car was produced by DKW, and featured front-wheel drive and a dynastart. Despite the economic troubles, demand exceeded supply. Auto Union produced a number of stunning racing cars, with a V16 engine developing 520hp. Zwickau's engineers were responsible for much of the design work. By 1939 Auto Union had acquired eighteen Grand Prix records. Horch stopped producing cars in the Second World War and specialised in armoured personnel carriers. Audi made lorries. Forced labour was used until the Zwickau factory was bombed in November 1944. After the confiscation of much machinery and documents, a small repair shop was set up in the ruins of each of the two Zwickau factories to refurbish Soviet vehicles. The highly skilled engineers also produced agricultural equipment, but lacked the tooling to produce any cars. The East German Communist Party nationalised any business belonging to Nazis and war criminals, but they were soon none too choosy about who the businesses belonged to, and Auto Union passed into the 'hands of the people'. Vehicle production restarted in 1947 with the H3 lorry at the Horch factory. It was made of pre-war materials and could carry 3 tons. The V6 engine developed 110bhp. In 1948 the old names were swept away, and the Industrial Vehicle Building Association (Industrieverband Fahrzeugbau or IFA) was established. Horch suffered the indignity of being required to produce tractors, and the first Pioneer tractor was handed over on 21 May 1949. The company continued to produce heavy lorries while Framo and Phanomen produced lighter lorries. Horch handed over lorry production in 1959 to the Ernest Grube Works at Werdau. History does not relate who Grube was, although chances are he was some minor political functionary.

Car production resumed at the Audi factory with the IFA F8, pictured below, which corresponded to the pre-war DKW F8.

The F8 had a 2-cylinder two-stroke water-cooled engine of 690cc capacity, and three forward gears. Because of the lack of steel, the bodywork was of wood with artificial leather covering. Between 1949 and 1955, 26,254 were produced, while the replacement model, the F9 (seen here), was jointly produced by Horch who produced the bodywork and Audi who assembled the vehicle. The F9 had a 3-cylinder two-stroke engine that developed 28hp from 900cc. Top speed was 68mph. From 1949 to 1952, 1,625 vehicles were produced; then production was shifted to Eisenach, where the F9 formed the basis for the Wartburg. East German politicians lacked a limousine and the pre-war 6-cylinder 2407cc Horch engine was used in the P240 (below). This luxurious machine boasted reclining seats, separate heating for driver and passengers, continuously variable instrument illumination, anti-dazzle rear view mirror and reading lamps. Top speed was 87mph and the P240 weighed 3,256lb. Production was limited, with only 1,382 examples being produced between 1954 and 1959.

Since 1951 an IFA research committee had been attempting to produce a material for car bodies that was cheaper and lighter than steel. Eventually the group came up with duroplast, which was based on phenolic resin reinforced with cotton. The first duroplast car was the P70, essentially an IFA F8 with an up-to-date body. The P70 was produced between 1955 and 1959 in the form of saloons, estates and coupés. A total of 36,063 vehicles were built. Nissan's recent Figaro borrowed heavily from the styling for the P70 coupé. The chassis of many P70s have rusted away, but the cars that survive are highly sought after. The dynastart worked very efficiently, unlike the heater, fed from a heat exchanger over the exhaust. The fuel tank contained 7 gallons, which was good for 250 miles while two-stroke oil was added at a ratio of 1:25. Later cars used a leaner mixture thanks to better bearings. According to the owners' manual, duroplast had been used for the body for the first time in series production. The new material had important advantages over metal, including low weight, good heat insulation, good vibration/silencing properties, weather resistance, and an extraordinary degree of elasticity. Even long journeys in a P70, amid its leather trim, are pleasant and relaxing. The saloon and the coupé are both pictured here.

While the P70 was small, it was still expensive to produce and a new derivative, the P50, was designed – slightly smaller and simpler. Employees were asked to think of a suitable name and these ranged from Muldeperle (the local river) to Robert Schumann car (as Schumann was born in Zwickau). Eventually the panel chose Trabant. The first Trabi came off the production line on 7 November 1957 and fifty cars had been produced by the end of the year. Water-cooling was replaced by air-cooling driven by a V belt from the camshaft: very simple and effective, given that the faster the engine turned the more air was blown over the cylinders. Maximum speed was 55mph from 500cc and a compression ratio of 6.7:1.

In order to increase production the Audi and Horch factories were required to merge, and on 1 May 1958 VEB Sachsenring (formerly Horch) and VEB Automobilwerke Zwickau (formerly Audi) became the VEB Sachsenring Automobilwerke Zwickau. The two factories were still separated geographically which meant a great deal of trundling parts between the two units. For the early 1960s the Trabant was not a bad proposition, and the car was exported to Holland, Belgium, Denmark and West Germany. Between 1958 and October 1962, 130,000 P50s came off the production line. Improvements to the P50 included a new instrument panel on the P50/1, while on the P50/2 the gearbox was fully synchromesh.

The first P60 was launched in 1962 with a 600cc engine and 23bhp (compared with 18bhp). The P60 (also known as the Trabant 600) was built between 1962 and 1964, and production totalled 150,000.

For twenty-six years from 1964 the 600's successor, the 601, was built with barely any changes. The development budget for the 601 was virtually zero but Dr Werner Lang, the chief design engineer, did a remarkable job of producing a roomier and more modern vehicle. Ventilation was improved by wind-down rather than sliding windows, while a useful parcel shelf was fitted under the front dash. Clever plastic wind deflectors in the rear upper part of the windows provided draught-free ventilation, and these survived until the final Trabi came off the production line. The radiator grille could be removed by loosening two wing nuts; this provided better access to the carburettor. The bonnet was fitted with an internal sound insulation mat while the air deflector, which takes the cooling air downwards behind the cylinders, was coated with an anti-noise filler. The whole of the floor of the car including the boot was fitted with a sound insulation mat. The front silencer had a better heat exchanger and an additional petrol heater was offered as an option. Brakes were now self-adjusting and a centrifugal regulator improved cold starting. This curious device often confuses the man from the AA, and likewise DIY repairers when they reset the points. East German motor magazine editors had very little to write about, so every improvement was reported as a masterpiece. Those in the know found a suitable helping of sarcasm or irony between the lines.

The saloon was pompously known as the limousine, while the estate car (opposite, below) was known as the universal. The deluxe version had a different colour roof, chrome bumpers and rear demist. In 1965 an automatic version, the hycomat, was launched. This comprised a hydroelectric clutch system and was the only automatic available in the Soviet bloc. A further 3hp was squeezed out of the engine in 1969. After fifteen years the one millionth car had been produced, and gradually a number of improvements were made – such as the camshaft acquiring needle bearings, which made it possible to reduce the oil mixture to 1:50. Dual circuit brakes were introduced, and in 1983 a 12 volt system with alternator. A Jeep version was produced for the army, and in order to further economise on fuel the accelerator pedal was given a point of resistance three-quarters of the way down. A fuel consumption indicator, whose light diodes showed green, amber and red according to fuel flow, was introduced – and soon nicknamed the 'mouse cinema'. In the mid-'80s the rear leaf springs were replaced with coil springs.

Today Trabis are still popular in Germany, and there is even a cabriolet version available (above), made by Osterman of Osnabrück. The Trabi enjoyed a surprising number of rally placings, in its own class of course. The engines were tuned up to 70 or 80hp and top speeds were over 105mph. Trabants entered the Monte Carlo Rally, the 1,000 Lakes Rally in Finland, the Acropolis Rally in Greece and the Lombard RAC in the UK.

When VW took over the Trabi factory they found working conditions completely alien to the West German pre-occupation with health and safety. Firstly many of the workers were North Korean, flown in on the once-a-week Air Koryu IL62 flight from Hanoi to Berlin. Ten years after Trabi production stopped, the flight is still listed in international timetables, although most weeks the flight is cancelled with no explanation. Remnants of the Horch logo can still be seen on the factory brickwork, and the VW video of production shows endless spot welds being made by hand. The duroplast production unit was quite a sight, with every piece of kit covered with a fine white dust. Duroplast is a superb substance, being light yet strong, yet it was ultimately the factory's undoing: every duroplast panel had to cure within the press, resulting in huge bottlenecks. The engines were virtually handbuilt (by German craftsmen not Korean). The duroplast panels were hand finished, using some very nasty looking band saws with no protective guards. The panels were then glued and screwed on to the body and then painted by workers often without masks. Every vehicle went through a water test, which comprised a deluge of water from above while a worker with a torch examined every nook and crevice of the car. Finally the cars were driven at breakneck speed on to car/train transporters, and as the crossply types screamed with protest one of the rear wheels would lift, as if in final salute to the factory.

Dr Werner Lang was behind most of the Trabi prototypes, although he was actively discouraged from devoting time to development work by the GDR government. The P603 (above) featured a four-stroke engine, automatic transmission and fastback body. The P610 (below) was another prototype developed to a full roadworthy condition. However, the politburo blocked any funding. There was a fear that the prototypes would be destroyed for political reasons so they were hidden in a cellar, and can now be seen in the Horch museum in Zwickau.

Throughout the 1980s politburo members discussed *ad nauseam* how the Trabi was to be developed. Little happened until the death of Leonid Brezhnev, when cautious criticism became acceptable in the USSR. Although at first the GDR banned a number of Soviet publications for being too critical, a programme was finally agreed to modernise the whole GDR motor industry between 1987 and 1990. A 1.1-litre engine for the Trabant was agreed upon in 1984, but the first Trabant 1.1 with a VW Polo four-stroke engine built under licence only surfaced in autumn 1989. In November 1989 two strokes and four strokes were being built in tandem, and on 21 May 1990 the three millionth Trabant, a 1.1, came off the production line in the new Mosel plant. The first VW Polo was assembled from components there at the same time.

On 26 June 1990 mass production of Trabants ceased, although a few Tramp leisure vehicles, like this one, were produced until 30 April 1991, when car production finally came to a stop. Trabi mania continues in Germany, with huge summer rallies and a regular magazine *Super Trabi*. Trabis have been shortened and stretched, and a professional conversion kit is available for turning saloons into convertibles. Relatively recently a consignment of four-stroke Trabis which had been sent to Turkey but never sold were shipped back to Germany and remarketed. There are still a very limited number of two-stroke Tramps for sale at around £2,000. These cannot legally be registered in Germany because of emissions regulations.

Eisenach had started producing Wartburg cars as early as 1898, and spent much of the first two decades of the twentieth century producing large cars with a limited market. The early Wartburgs were a very high-quality product and, despite modest sales, won much praise. The economic collapse which befell post-war Germany naturally hit the works very hard, and a number of spells ensued when car production ceased. Eventually BMW, seeking a base from which to expand from its aero and motorcycle business into more diverse markets, stepped in and bought the factory. Its plan was to build a car capable of mass-market appeal. Like so many manufacturers before and since, it elected to use a tried and tested design from another company. Following negotiations with Austin of England, production began in 1928 of the BMW Dixi, a copy of the Austin Seven small car. The Wartburg name was revived during 1932 for a particular BMW model, and with assistance from the German state development started on a range of large, 6-cylinder cars. Of great elegance, quality and sophistication, the original 1930s 3-series BMWs not only sold well in the improving economic climate, but also gained the firm a reputation for building solid, reliable and highly presentable cars. Naturally the Eisenach works contributed in no small part to the German war effort, building cars, motorcycles and other goods for the army. It was thus a priority target for Allied bombing efforts, and was 60 per cent destroyed during the Second World War.

After the war the medieval town of Eisenach lay in the Soviet sector. Much of the remaining German industrial base was dismantled and transported to Russia as part of war reparations. Entire railway lines were torn up and used in the Soviet Union; whole factories disappeared east. It looked as if a similar fate would befall the BMW works. It was, immediately after the war, producing wheeled vehicles such as wheelbarrows and hand carts when an order went out that any factory which did not wish to be dismantled and closed down should show the articles it was capable of producing. Desperate for the future of their town's economic mainstay, the workers of Eisenach scraped together enough spare parts, new and secondhand, to put together a few cars. Upon inspection, it was evident to the authorities that here was a going concern (a magnificent act of deception on the part of the factory management), and funds were approved to begin recommissioning the factory. From this venue was born the 'Avtovelo BMW' – a copy of the authentic BMW in almost every detail. Production restarted, to howls of protest from BMW, especially when soundings were taken regarding exports. After pressure was brought to bear the factory's product was renamed EMW – Eisenacher Motorenwerk. The product was now different to its forebear in the small matter of its badge. The four-quadrant white and blue badge became cream and red, and the letter B had its ends filled in in the production dies so that the badge now read EMW rather than BMW. The factory itself meanwhile continued with its historic name of Automobilwerk Eisenach, or AWE, a set of initials still cast into certain components at the very end of production in 1991.

Production of the large, 6-cylinder BMW models continued, with developments, into the 1950s. But economic planners of the GDR recognised that this expensive, limited production car would not provide mobility on a larger scale. Virtually all cars had gone to government departments, and as a private car its expense almost eliminated it as an aspiration of any but the high elite of the fledging state. At Zwickau, some 60 miles away, however, a better prospect was in production. The former Horch works had been producing very small numbers of the pre-war designs of their big luxury car. It had also been making copies of pre-war DKW front-wheel drive automobiles. These small, innovative and relatively cheap cars were a modern, mass-market product. Whilst Zwickau was to continue producing the very smallest of these cars – later to evolve into the worthy, but eventually anachronistic Trabant – the medium car, the IFA F9, was to replace the EMW at Eisenach.

It would be unfair to say that the IFA (Industrieverband Fahrzeugbau der DDR – Industrial Vehicle Building Association of the GDR) F9 was an exact copy of the post-war West German DKW F9. The East Germans, to their credit, had built prototypes and discovered a number of shortcomings in the design. This had not happened at DKW, so although the two cars appeared identical the IFA was in fact a superior product, with many bugs ironed out – some while before numerous failures led DKW to modify its production. At first the F9 was a shock to the workers at Eisenach. To some extent they were craftsmen, accustomed to the niceties of design and detail that a luxury car affords. The F9 was built to a price as well as to a specification and was relatively crude. Its engine had half the number of cylinders of the BMW and only seven moving parts. Its interior was spartan as an economy car always is. Increases in vehicle production led to less time for attention to detail. Still, as time went on the Eisenach Motor Works became as proud of its new product as it was of its old. The major morale boost came with the first car since 1932 to bear the hallowed Wartburg name.

The Wartburg 311 of 1955/56 was quite similar to the IFA F9 in its main mechanical units. It had been developed and refined in many ways, but the unmistakable DKW underpinnings were all there. It was set apart from its earlier incarnation by an all-new bodyshell. Of elegant, low and flowing design, the Wartburg 311 was a truly beautiful car, which still attracts admirers the world over.

Deutsche Strassenverkehr, the GDR's most popular motoring magazine, naturally gushed with praise. The workers glowed with pride. Export markets opened up. The 311 saloon was developed heavily in its first year; then, once the mechanicals were proven, the designers turned their hands to ever more exotic variations. The 311 station wagon was a plain yet roomy estate in a two-door incarnation (the saloon always had four doors). Sweeping coupés followed with wrap-around rear window glass. A two-seater coupé appeared with a short passenger compartment but huge bonnet and boot – the entire engine and gearbox (longitudinally mounted) were accommodated in front of the bulkhead.

Ever more elegant convertibles followed, both as stunning production types and amazing prototypes. Two-tone paint adorned many models.

The most extravagant estate car was the Wartburg Camping (built by Glaser), whose slab sides rose from a standard estate car's side panels to a crescendo of curving clear perspex cutting into a high steel roof. It remains unique. Its 6ft 2in loadspace meant that if all else failed on a camping expedition, comfortable beds could be made in its rear area.

The presence of many 311s still in regular use throughout the world is testament to their durability, but the process of slow development continued. Its transverse leaf springs front and rear give it unparalleled cornering abilities, outperforming many modern cars on the B-roads, but with this set up comes a certain harshness of ride. Furthermore, driveshaft life is not unlimited. The 312 of 1964 used an externally identical bodyshell, but with a completely new chassis and gearbox. It was one of the first mass-produced cars to adopt fully independent coil spring suspension. Service intervals rose to 30,000 miles – unprecedented in its time – and the weaknesses of the previous transmission were overcome. The basic design of the driveshafts is still seen in today's VW and Audi cars. Thirteen inch wheels were adopted, in line with the modern trends of the time, and the Wartburg 312 was, mechanically at least, ahead of its era. Designers slaved away on two- and four-stroke monocoque prototypes, but limited development finances led to the further evolution of the 311/312 design. One prototype even used a glassfibre sports body with a flat, mid-mounted engine driving the rear wheels.

As 1964 saw a new chassis under an old body, 1967 saw a new body to clothe this chassis. To my eyes, model 353 (shown here) is a crude, boxy design, with none of the elegance of earlier cars. However, in context, it was a modern three-box angular affair which reflected the trends of the mid- to late '60s. However one looks at it, it is, at least in appearance, a rough contrivance, with huge panel gaps and a bonnet and boot lid which might have been made for a car that was an inch smaller all round. Safety played a part in the design. A collapsible steering column was incorporated, and all controls except for rotary lighting and windscreen wiper switches were placed under the dash, so as not to cause injury in an accident. Panels such as the inner wing assemblies bolt on to the bodyshell for easy replacement, and rear wings also unbolt for ease of repair and maintenance. Contemporary road testers viewed its ride as far superior on soft coil springs, though they fairly commented on its large body roll, exaggerated by special rough-road shock absorbers which have no resistance on the upstroke, so as to quickly retract a wheel that has hit a rock or pothole. *Motor* magazine tested an example fitted with radial ply tyres and two-way Woodhead shock absorbers, and found it a great improvement. A huge estate car, the Tourist, was more capable of swallowing loads than much larger cars and, ever innovative, incorporated a glassfibre tailgate and rear wings. Here, development slowed almost to a halt. Occasional tinkerings with engine port timings and, in 1974, a new crankshaft design which increased crankcase compression, brought power gains (at the expense of low revs torque), but otherwise the cars remained unchanged until 1976, when twin circuit disc/drum brakes were adopted. For some years rally cars had been fitted with disc brakes from, of all things, the Austin Maxi, but the new set-up used four-pot calipers from Skoda – a couple of years, incidentally, before Skoda itself incorporated these into its top of the range cars.

This use of Skoda parts was the sad remains of once great intentions to develop a range of cars jointly based on East German and Czech technology. One of these designs from 1975 used a bodyshell which bears a remarkable resemblance to the Skoda range of 1978.

No more changes were made until 1984/85, when the Wartburg 353 got a revised dashboard with electronic LED instruments. The most entertaining of these was a twenty segment fuel flow indicator quickly christened 'Das Mauserkino' – the Mickey Mouse cinema. More significant was the removal of the small brass radiator mounted on the chassis behind the engine. This delightful anachronism was replaced with a crossflow aluminium Skoda radiator of large size, squeezed between the front of the engine and a slightly extended front panel. This panel was now a one-piece pressing, painted body colour, with squarer, larger headlamps. A brief note is now required on the subject of headlamps. These must surely constitute East Germany's biggest sales success since Praktica cameras. Shortly after they appeared on the 1967 Wartburg 353, they started to be exported from VEB Fahrzeugelektrik Ruhla (the main East German transport electrics factory) to Moscow, where they were put into the fronts of Moskvich 412 cars until 1991. Flush with the success of this idea, the Soviet authorities designed them into the Latvia van range, PAZ buses and endless passenger and commercial vehicles. An entire fleet of police Ural motorcycles was constructed with an ungainly, oversized Cyclops Wartburg eye beaming out from their distended fairings. Whenever trawling through photographs of Soviet vehicles and prototypes, it is always possible to spot the Wart headlamps. They are always there. One suspects the designers had little choice in the matter. . . .

Back to Eisenach, and the 'new' Wartburg 353 with its front-mounted Skoda radiator acquired a Czech Jikov twin-choke carburettor, of far more complex design than the crude (but effective) Berliner Vergaser Fabrik (Berlin Carburettor Factory) item which had graced it for so long. Whether all these improvements were necessary is highly debatable, but they did at least provide some means of distinguishing between the 1985 model and that of 1967. The real problem, however, was that the two-stroke engine was becoming less and less politically correct all over the world. It is possible to sustain a good argument that the infinitely repairable, reasonably economical two-stroke Wartburg is a very environmentally efficient car in many respects. However, more and more countries were legislating against two-stroke use in motor cars. Export markets dried up, one by one. Britain saw its last imported car in 1974/75, with stocks sold until 1976; but by the mid-'80s even fellow Comecon countries were becoming increasingly reluctant to take new Wartburgs and Trabants.

It is a little known fact that for many years the engines for Volkswagen's small car ranges were made in the GDR. Had buyers been aware, they might have stayed away in droves, such is the prejudice against Communist bloc products in Britain. However, as we all know, Volkswagen's small engines are remarkable in terms of quality and longevity. It is little surprise, then, that permission was sought to use these engines in Wartburg and Trabant cars. Although the cars are known as VW-Wartburgs, the engines are in fact subtly different in many ways to the VW originals and nowhere bear the VW logo; instead, they wear IFA initials.

The first VW-engined cars came on stream in 1988, produced alongside the two-strokes. The form of the car is very familiar, a 353 with a nip and tuck, but in fact the car was comprehensively re-engineered, both mechanically and stylistically. The front portion of the chassis was widened, which allowed the new engine to be transversely mounted with its new gearbox. The front and rear tracks were widened, bringing a transformation in handling. A centre console interrupted the flat floor and a new dashboard updated the interior, with improved trimming throughout. ABS bumpers were added, with flared wheelarches, and flush indicators replaced the orange 'ears' screwed to the corners of the two-strokes. Flush rear lamps replaced the projecting afterthoughts of the original. Altogether, the car was a modern variant on a classic theme. Sales surged and there was talk of reopening export markets long closed.

1989 saw the end of two-stroke production at Eisenach; the close of a near forty-year era where the GDR had come, by default, to dominate the world of two-stroke cars. Now the four-stroke Wartburg 1.3 came on full stream.

Unfortunately political events hit the new wunderkind head on. The opening up of the GDR in 1989 meant that many East Germans were dazzled by more modern, yet often inferior cars from the west. Bored with their Wartburgs and Trabants, they clamoured for diverse products from West Germany and beyond, and large numbers of the cars with many years' service left in them were scrapped or abandoned. The failure of these imported novelties, given their often great age and dubious condition, led to GDR cars being de-mothballed after a few months, but the writing was on the wall. Sales of the previously very popular Wartburg 1.3 slumped after a surge in 1990, and even attempts to dress them up in modern bodykits failed to shift them in any great number.

Cooperation agreements with Volkswagen were mysteriously terminated and the Treuhand body, set up to dispose of the East German industrial complexes, handed control to Adam Opel GmbH. The factory was promptly shut down in 1991. Opel promised a bright future for the workers in its new plant to be built at Eisenach West, but ultimately on 1,500 people were to be employed there, in contrast to the 5,000 who had found work at Automobilwerk Eisenach. The works was razed to the ground. Only the factory gateway still stands – saved by popular demand after it became evident that this last remnant of one of the cradles of the automobile industry in Germany was also to be 'disappeared' along with so much else that represented the GDR.

Today the early Wartburg models of the 1950s and 1960s are much prized in Germany, East and West, but the square, utilitarian 353 is unloved and fetches virtually no money unless it is in concours condition. The Wartburg 1.3 is also a cheap buy, but this at least can be fitted with a catalyst to avoid punitive German taxation banding for older cars. It is a sad end for a long tradition of motor building, which survived against the odds on many occasions. It is also sad that so many infinitely repairable, strong and durable cars should have been put to scrap so long before the end of their natural working lives. When a waiting list for a car is eighteen years, to build it so that at this age there is still much life left in it is a great achievement. That it should be purposely designed this way is almost an act of philanthropy to those of us who respect sensible and expedient engineering.

Wartburgs were exported to many countries over the years. Europeans took to them and they remain a popular car in Greece. One batch for Egypt was delivered in 1978. Owing to a spat between the Egyptians and the GDR at the time, the cars were effectively abandoned at Alexandria docks for some three years. The baking sun bleached out and rotted the upholstery, vandals took pot shots with stones and Egyptian dock workers ran them around on oil-less petrol, seizing many engines. In 1981 the cars appeared on the dockside at Sheerness in the UK. Exactly how they came to be here remains a mystery that no Wartburg aficionado has yet solved. Here they sat, salt water now attacking the sun-dulled paint and shrunken window seals until a dealer in Lancashire was approached by Wartburg GB, asking if they wished to purchase the vehicles as spare parts. They were all left-hand drive and by now had no type approval for sale in the UK. This did not daunt them. Buying the cars for £360 each, fifteen were dismantled at the docks and the remaining 105 transported north. Using both stock and secondhand parts for steering, dashboard and headlamps, they began to convert the cars to right-hand drive and sell them to customers who had enjoyed the charm of past Wartburgs. To get round the type approval regulations, they claimed officially that all the cars were stripped down, put into the parts store and 'new' cars built up from this store. In reality, bodies were removed from chassis, conversion work undertaken and then, reassembled, the cars were sold on with a new Y registration (for 1982/83) as 'Wartburg Lancastrians'. After twenty-five were sold, the local DVLA unfortunately realised that a complete strip and rebuild must be impossible at the price (£1,500 for the saloon and £2,400 for the estate) and gave notice it would refuse to register any more. The dealer's ingenuity triumphed, however. The last sixty-five cars were sold to owners of old Wartburgs, who drove their wrecks to the garage in the morning and went away for the day. The little elves swapped chassis plates from the old car on to a new one, transferred the number plates, and by 5.00 p.m. the happy owner could leave with a new Wartburg on an ancient registration.

The unconventionality of the Wartburg strikes you as soon as you attempt to enter the car. The locks are not spring loaded and the key remains in the unlocked position as you withdraw it. Closing the door results in a resonant clang. The interior of the car will be familiar to anyone who has travelled on an East German tram. Metal fixtures abound, finished in silver paint, and moulded plastic is absent. Turning the ignition key in the lock gives a satisfying clunk, and the instant that the starter turns, the engine bursts into life with a sort of distant zinging, whizzing sound.

Casting an eye over the dash reveals a two-dial instrument panel – a large speedometer with a similar unit next to it containing gauges labelled 'temp' and 'tank'. They are separated by twin columns of warning lights, many of which are unused, waiting patiently for the next specification upgrade. Most controls are hidden under the dashboard, a noted safety feature at the time, with only the switches for lights and wipers obvious to casual inspection. A slightly home-made appearance to parts of the dash belies the special origins of this particular car: it's a 'Lancastrian', built in 1978.

I pushed the car from a garden in 1995, where it had sat for a few years with a seized water pump. Its rust and algae-stained yellow paintwork was drab and unimpressive, and it looked like a good spares donor for the running estate car which I was slowly restoring. By 1997 it was back on the road again.

Back to our test drive. Setting off down the street, the first impression is of the great eagerness of the engine, which, although only of 992cc and 50bhp, propels the car from rest with great gusto. Snatching second gear, the synchromesh gives warning that it will not be hurried – it works well, but slowly – and the engine emits a very high-pitched sound, unfamiliar to any four-stroke car owner. As we near the bottom of the street, we must abandon the throttle and apply the brakes. Here another special feature comes into play. As the accelerator pedal is released, the engine drops to a slow, contented burble and the car cruises onwards in eerie silence. Inside the gearbox lives a freewheel, which is in principle like a bicycle freewheel. If the engine speed is slower than required to propel the car, it disengages and the car coasts. This feature is necessary to spare the two-stroke engine from the wear and tear caused by over-running with the throttle closed. In a two-stroke, no throttle equals no lubrication. It also removes a great deal of driveline shunt caused by the engine kicking and surging, which is a consequence of the ignition timing being fixed at 22 degrees btdc – far too advanced for smooth over-running and also the reason that the engine splutters and burbles at idle.

As we coast towards the end of the street, we can apply the disc/drum brakes. Unservoed, they are none the less firm and powerful – powerful enough to lock the front or back wheels without much effort. We slide the enormously long and spindly gearlever into neutral without the clutch, and at walking pace feather the throttle and ease the car back into first – again without the clutch – and set off up the main road.

Let's get a move on. Clutchless gearchanges up the box are fine if you let the revs die and gently press the lever home as the synchro gives way. However, for more pacey progress we use the clutch as per normal. Pressing the throttle home gives comparatively meteoric acceleration on engines made after 1974. The engineers at Wartburg were always tinkering with cylinder port timing to alter performance. Around 13 seconds to 60mph is normal, not bad from 50bhp and a ton kerb weight. This car, however, has had an early '60s engine installed. The joy of this is that there is endless torque at low revs and it will rumble along at 1500rpm or less. This is always a source of amazement to motorcyclists, whose highly strung two-stroke units are completely below screaming pitch of 6000rpm or more. It is well suited to the Blackdown Hills of Devon and Somerset, amongst which I live.

To make swift progress, we hang on to the revs, though, changing up at 4250rpm when this engine makes its maximum power. Characteristic of these older engines is that they give out a pseudo-turbocharged whistle at full load. Most pleasing to the ear, and fair warning of the likely fuel consumption!

At the end of the journey, the engine can either be turned off with the key, when it will still burst into life again without throttle some hours after being parked, or, if it is to be left for longer, it is strangled with the choke. This floods the cylinders with neat vapour, again giving instantaneous starting when next required – even after a week of disuse.

The final quirk of the car is that it cannot be simply left in gear: the minimal friction inside the engine means it will creep forward on the slightest slope, and so the handbrake – silver painted, of course – must be applied.

Perhaps the nicest thing about this car is the amount of interest it provokes. Its brash paintwork, blunt, crude styling and its unique engine sound means that heads turn wherever it goes.

The Barkus B1000 is East Germany's answer to the Ford Transit, and is Wart powered. This particular camper was imported from the Czech Republic and is shown in London, waiting to be filmed for *Top Gear*. Note the British Tatra 613 in the background.

This quaint period picture shows the Framo as being larger than it really is. The Framo was produced in the '50s as East Germany's standard van.

6

Hungary

Mention Hungarian auto production to a Western classic car enthusiast and you are likely to get a blank look. Hungary is better known for its Ikarus buses, which pop up in the most unusual of places. For many years Istanbul's public transport system was purely Ikarus driven, and given the state of Istanbul's roads this says a lot for the buses' durability if not their comfort. Ikarus has produced buses for many years and some of the earlier models are quite attractive. Imagine if you can a bus with Tatra styling motifs and you have the Ikarus 55.

There is more to Hungary than simply bus production, and while the Soviets decided that the Hungarians did not have the necessary automotive skills following their occupation after the Second World War, the Hungarians thought otherwise: Hungary has produced some extremely weird microcars. When the Second World War ended Hungary was ruined, and so was its transportation. Motorcycle production started fairly quickly (at the Csepel motorcycle plant) and most of Hungary's microcars used the Csepel engine. Between the wars a number of Hungarians experimented with light cars, and these skills were used after the Second World War.

One of the first projects was the Suranyi 125 (pictured here), handbuilt by an engineer of the same name. The dimensions are pure pedal car and only two were produced, using motorbike engines, one with 125cc and the other 250cc. A 1 in 20 incline was enough to bring the original 50cc machine to a halt.

By 1952/53 the Hungarian authorities were trying to figure out how to get around the prohibition of Comecon on the production of passenger cars. They focused on closed four-wheel motorbikes transporting two adults and two children or luggage. The Ministry of Light Industry was in the forefront, and with their help the Úttörö (pioneer) microcar was born in Debrecen, 120 miles east of Budapest in 1954. It was designed by Janos Schadek who was an engineer student in the '20s (he even built a Morgan-like tricar as his first project). After the war his family businesses were nationalised and he found himself Chief Engineer of the Locksmith and Iron Foundry Company. Two friends of his repaired and assembled aircraft, and the three men got together to build the Úttörö from scratch. It was powered by a Csepel 250cc motorbike engine and there was a kickstart. The wheels were taken from a barrow. Early steering (directed straight to the front axle) was oversensitive; thus rack and pinion steering was substituted. The car reached a maximum speed of 50mph.

The Szekesfehervar Motorjavito Vallalat repaired aeroplanes before the war. When the aeroplane business was shut down they looked for new activities, and using scrap aluminium they started to make everything from antennae to cheese-making equipment. Microcars followed shortly after, although because of Comecon's prohibition of car production they were registered as invalid vehicles or motorbikes.

By early 1956 two microcars were ready: the Alba Regia (above) and the Balaton (below). The power unit was the 250cc Pannonia motorcycle engine with an output of 8bhp. This was no match for the 14bhp from the Goggo. Transmission was borrowed from the BMW Isetta, reverse from the Messerschmitt. When the driver wanted to go backwards he stopped the engine, pushed a button and started the engine in the opposite direction. Wheels were aircraft tail wheels.

The doors of the Alba Regia opened conventionally but the Balaton had a swinging back door just like an aircraft. An engineer from the Csepel motorbike factory designed a unique pressed steel-framed bike with torsion suspension, and this system was incorporated in the Alba Regia. The Balaton received a strange suspension: the arms were held by rubber tags with vibration being absorbed by the rubber.

Built near the town of Balaton was an unnamed bug-eyed microcar, with an NSU 500cc engine. The steering is by aircraft control column, because of a lack of space above the driver's knees.

In the May Day parade of 1956 four microcars bounced along the streets of Szekesfehervar – namely the Úttörö, Isetta and the locally made Alba Regia and Balaton. Even though the Goggo was clearly the more refined vehicle, the Ministry decided in August 1956 to go ahead with manufacturing plans. However, in late October the Soviet troops rolled into Hungary and the Russians ordered the Hungarians back to their mopeds. The bubble car was slowly being forgotten.

The Gitta, illustrated here, was a project that fell by the wayside.

The 1956 Revolution in Hungary put an official end to the Hungarian microcar efforts, such as the Alba Regia, Balaton and the Úttörö. But individual efforts still surfaced from time to time: Kálmán Szabadi's was one of them. He had a colourful life and his car reflects this. It was powered by a 300cc Isetta unit and was completed by 1960. Resin, which wasn't available, was replaced by mixing pig fat and chicken feathers with nitro shellac and adding a layer of paint. One has to admit that the smell coming out of the car was, hmmm, interesting. Entry and exit was provided via gullwing and conventional doors, as there wasn't another way for our over 6ft tall hero to get in and out of the car. Shock absorbers were Pannonia motorcycle units, but as they were known to crack easily Szabadi employed two of them for each wheel. The car looks big on the pictures but in reality it was only 10ft 2in long and weighed just 836lb. Here we have to pay tribute to an Hungarian automotive journalist. Gyorgy Liener saw the finished car and, although the smell might have distracted him, he didn't hesitate to contact the officials and try to convince them to produce the car. But there was no way to avoid the Comecon regulations and the car remained a one-off.

7

Romania

There was no mass production in Romania before the Second World War, although a French company produced some 150 cars at a plant in Arad from 1910. The most remarkable prototype to come out of Romania was the Persu Streamliner, built in 1920. This teardrop-shaped machine had wheels within the bodywork rather than under mudguards and was constructed by Aurel Persu, even receiving a German patent for its streamlining.

The first attempt at a Romanian people's car was the 1945 Malaxa, built by engineers from Malaxa, Bucharest and IAR Brasov led by Petre Carp. In true VW style, the engine was at the rear but a 3-cylinder radial was used. Top speed was 75mph and fuel consumption was 28mpg. It also proved to be the last attempt to build a Romanian's people's car.

The origins of Aro (Automobile Romanesc) are shrouded in mystery. It would appear that the first Aro, known as the IMS 57, was produced in 1957 and this four-wheel drive machine would appear to be very closely based on the GAZ 69. Quite possibly the Soviets passed over the tooling while continuing GAZ 69 production back home. A 3260cc engine was mated to a chassis in an old military plant in Campulung, which formerly produced 'shooting equipment for aeroplanes'. The rolling chassis was then tested and passed to the Pitesti plant for bodywork and final finishing. The large engine produced a pathetic 50bhp at 2800rpm, so two years later an upgraded machine was produced, the M59 with a 56bhp engine. The top speed crept up from 50mph to 56mph and manual windscreen wipers were replaced with electric ones. The rather odd trek to Pitesti for finishing ceased and the machines were completed in Campulung. There was a choice of two doors, four doors or pickup, and productivity rose to a heady 3,222 vehicles in 1963.

The M461 was launched in 1964 and, while still based on the GAZ 69, had a 70bhp engine with a maximum speed of 60mph. Exports started to China and Columbia and the M461 won a number of competitions including the 1970 Forest Rally in Belgium. Production of the M461 totalled 80,233, of which over half were exported. Production ended in 1975. The next Aro was a purely Romanian design, and in true Communist style was born after a long gestation period, work starting in 1966. The Aro 24 series came in a variety of versions, with two doors (240), with four doors (244) and in station wagon form. Pick-ups (the 242) were either in long wheelbase or short wheelbase, while there was also a three-door hard top (243) with a soft-top option. For its time the Aro 24 was not a bad machine and, with coil springing up front, the ride was better than a number of Western counterparts. During a test organised by Ford Europe in 1972, the Aro 240 came second after Range Rover but before Jeep, Ford or Land Rover. A 2495cc (83bhp) petrol engine or 3119cc (68bhp) Brasov diesel engine were available. The rough terrain characteristics were impressive, with the car able to climb slopes up to 70 per cent, tow trailers up to 2.3 tons and cross 2ft deep fords. Payload was 1,760lb. Exports were reasonably successful, but in order to increase its appeal vehicles were shipped to Portugal, where a Daihatsu diesel was inserted (2530cc) and while the power was only 79bhp at 3800 revs, reliability was much improved. The use of Renault 12 (Dacia) headlamps was perhaps a false economy but otherwise the styling was not unattractive. This version, the Portaro, sold quite well in the UK.

By 1985 over 125,000 Aro 24s had been produced of which 93,000 were exported. After 1989, to increase the Aro 24's appeal abroad, foreign engines were imported and a crewcab pickup and long-wheelbase estate were launched. There was even a Cosworth-powered luxury version, the Hunter. Right-hand drive Portaros were imported into Britain for a number of years; however, the survival rate was low. A specialist in Devon keeps the marque alive and has even persuaded aid convoys to Romania to bring him spare parts on their return journey. A plan surfaced to import Aros into the US in 1997, and with a sticker price of just under $13,000 quite a number of dealers signed up. Literature indicates a 3-litre V6 engine 'US built', but does not mention that it is a copy of a Ford/Cosworth V6 cobbled together by independent manufacturers from off-the-shelf parts. At the time of going to press problems remain such as the Aro's lack of airbag, and importation has not yet started.

This early publicity shot shows a Romanian trying to drive his Aro 24 into a rock face.

The Aro 10 (above) was launched in 1980 using the Renault 12-derived engine from the Dacia car. The styling can only be described as utilitarian, and once again the Renault 12 headlamps sit awkwardly in the grille. In 1984 the Aro 10 was fitted with the 1397cc Dacia engine which allowed a maximum speed of 73mph. Perhaps surprisingly, for the first five years of production 90 per cent of the vehicles were exported to Great Britain, Italy, France, Germany, Spain and Greece. The model was only known as the Dacia Duster in the UK (opposite); the French knew it as the Aro Trapeurs, while in Italy it was the Aro Ischia. Variants included diesel-powered versions (from VW and Pegaso), while the French also had the soft-top Spartana with Renault engines.

In Britain prices ranged from £4,594 for the pickup to £5,995 for the Duster GLX. Dusters were imported through Dacia Concessionaires, Westbury, Wiltshire, and agents often found owners coming back with cooked engines. Country Vehicle Services, in Silverton, Devon, is one of the few remaining specialists and, apart from fitting engines from scrap Renault 12s, they have carried out quite a number of conversions, using mostly diesel units. Aro continued in production in Arges, Romania, and while a 1989 prototype remains stillborn, 1999 saw a new 4×4 pickup prototype. The future looks grim, with demand falling, so much so that in the first quarter of 1999 total sales were 279, of which 31 were exported.

The name Oltcit derives from the Romanian word for country, 'olt', and Citroën. In the early '70s, prior to the Peugeot takeover, Citroën had been working on a new small car designed to replace the Ami 8. During the period of friendship with Fiat, this vehicle was known as Project Y. Following the Citroën/Fiat divorce, Project Y was scrapped and the unfortunately named Project VD ('*voiture diminuée*') was launched; this incorporated the floorpan of the Peugeot 104, and the end result was the LN, LNA and Visa. A new body was designed for Romania, and there are many Citroën styling cues – particularly the treatment of the rear wheelarch. Production started in 1982 in Craiova. The red regime insisted that 40 per cent of all components of the new Oltcit were to be sourced in Romania, since they were reluctant to spend hard currency on imported components. In addition, Citroën was forbidden to export hard currency profits, which is why it earned its return in kind: the Oltcit was sold as the Citroën Axel alongside other Citroën products in Europe. The Axel never sold particularly well, the Visa being far more popular. The economic disintegration of the Ceaucescu regime was only partly solved by the new government of Ion Ielescu (as most officials were old Communist hacks). Citroën withdrew in 1988 and the company was controlled by the government of Romania until 1994, when a new privatised company was set up, known as Oltena.

In 1996 the decision was taken to end the production of the Oltena, given a loss per car on sale of US$1,500. However, production may be continuing at Timisoara, where the producer of the ill-fated Lastun microcar (see p. 154) is considering restarting production. Two engines were available, namely a flat twin 652cc and an 1129cc. The 652cc is clearly the early Citroën Visa engine, a developed 2CV unit. The 1129cc derived from the Citroën GS was the unit fitted to the export version.

Various trim levels were available, with the base being TRE, followed by the RS and TRS, in true Citroën style. The minor controls were lifted from the Citroën Visa/GSA, but these may not have been retained in the Oltena. Prototypes included a cabrio and a four-door version named Oltina, which was shown at TIB 1989 (the Romanian car show) and is pictured opposite. There was even a wild Sbarro-designed estate. In 1994 Daewoo established a joint venture in the Oltcit works and in 1999 took 23 per cent of the Romanian car market. This is, however, outside the scope of this book.

With the change of name to Oltena, a pickup version was launched, the 12CS. A new front end included an air intake with heavy slats, not unlike those that adorned the late 1970s Talbots.

Dacia came into being in 1968 and has produced 2 million cars since. The first model was the Dacia 1100, essentially a Renault 8 built under licence. There were very few changes compared with the French version apart from a chrome plate labelled Pitesti Automobile Plant and an eagle logo placed between the headlamps. A limited production run of the 1100S was made for racing purposes, with quad headlamps, a heavy-duty battery and wider tyres. The water-cooled 4-cylinder rear engine produced 46bhp with a maximum speed of 82mph. Petrol consumption was 42mpg. The last Dacia 1100 rolled off the production line in 1971 after 37,546 cars had been produced. This particular example is Ceaucescu's at auction. Presented to him by the factory, it was sold off after his execution.

The Dacia 1300 was launched in August 1968, being shown simultaneously at Eren, Bucharest and at the Paris auto show. It was closely based on the Renault 12. The Renault 1289cc engine was manufactured in Romania, with just one or two small differences, producing 54bhp and a maximum speed of 90mph. Even in 1970 there were three alternative versions, the 1300 standard, the 1300 super and 1301, which was only available for comrades in the Romanian Communist Party.

In the last thirty years a huge number of variants have been produced, with, for example, the Dacia 1302 pickup being launched in 1973. This is a five-door hatchback.

The Model 1304 was launched in 1983: it was a pickup with single cab or crew cab.

In 1983 the old 1300 was lightly reskinned front and rear, and became known as the 1310 (above). The 1302 was launched in 1988, being a redesigned 1310 with the option of the 1397cc engine. The 1325 Liberta model emerged from the 1320 series and had a redesigned rear. Finally, the 1410 model is essentially the 1310 with the 1397cc engine.

The standard Dacia was considered infra dig for the Nomenklatura and they enjoyed the luxury of the locally assembled Renault 20, known as the Dacia 2000. Never exported was the Dacia Sport (sadly in name only). This two-door version of the Renault 12 is unique to Romania and was available with both the 1289cc and 1397cc engine.

The Nova is probably Romania's first home-designed car, and the long gestation period accounts for the slightly dated styling. A new 1557cc engine produces 72bhp with a 99mph top speed and 0–62mph acceleration of 15.9 seconds. Limited exports are being made, with Egypt being one of the customers. In 1998 Dacia took 88 per cent of the domestic market and made a profit, despite prices per car of around £2,500. The writing must be on the wall for the old Renault 12 variants following the signing of a deal with Renault in 1999. A new model is planned and production is set to rise from 10,600 in 1998 to 200,000 in 2010.

Romania's only attempt at a microcar was the Lastun, also known as the Dacia 500. It had a 2-cylinder air-cooled engine of 499cc. The power was 22.5bhp with petrol consumption being 85mpg the maximum speed was listed as 66mph. The Lastun was built in Timisoara, but production was abandoned in 1989. Considering the crudity of the design and frightful build quality apparent here this is hardly surprising.

8

Yugoslavia

Yugoslavia started car production late, although possibly the first four-wheeled Yugoslav machine was a Tam lorry. Production started in Maribor on 31 December 1946. Zastavas go back further than many people think, with the first agreement with Fiat being signed in 1954. With true Communist originality, Zastava means Red Flag. Production started in a former weapons factory in 1954, with the first model being the Fiat 1400, of which this is an example. The Campagnola was also produced – the Italian four-wheel drive vehicles.

Yugoslavia's most popular car was also its longest lived, and in 1955 production of the Fiat 600 started at the Kragujevach factory, with the line-up being extended in later years to the Fiat 750 and Fiat 850. The model range was quite broad, with restricted production of the 1100e, 1800b and the 2300. An astonishing 920,000 had been produced by August 1986 with the body style virtually unchanged in thirty years. Even the Fiat 600 Multipla was produced. No photograph from the Kosovan or Serbian conflicts would be complete without a Zastava 600, with its windows smashed, in the background.

As with Lada and FSO, the Yugoslavs produced variants not available in Italy. Their estate version of the Zastava 1300 was often used as a taxi. At the end of the '70s Zastava realised that it would have to update its model range, and the Zastava 1300 died at the end of the decade along with other long-in-the-tooth Fiat models.

ZASTAVA 1100 ZLM

MEDITERAN 3 DOOR HATCHBACK

Engine
Front, transverse mounted, 4 in-line: water cooled: 80 mm bore – 55.5 mm stroke: 1116 cc: 5 main bearing crankshaft: cast iron block: aluminium cylinder head: belt driven, overhead camshaft: compression ratio 9.2:1. Multijet, Weber style carburettor fed by mechanical pump. Maximum power: 55 bhp (DIN) at 6000 rpm. Maximum torque: 65lbft (DIN) at 3000 rpm. Sealed cooling system. Thermostatically controlled cooling fan.

Transmission
Diaphragm spring, single dry plate clutch. 4 speed (plus reverse) all-synchromesh gearbox with central floor mounted shift. Ratios: 1st 3.583: 2nd 2.235: 3rd 1.454: 4th 1.037: reverse 3.714. Final drive to front wheels 3.765.

Suspension
Front: independent by coil springs and MacPherson strut with an anti-roll bar. Rear: independent by self-stabilising transverse leaf spring with strut and trailing link.

Bodywork
All metal unitary body. Laminated windscreen. Toughened side and rear glass. Globocon PVC coating throughout underside, sills and valances. Plastic wheel arch liners. All body sections wax injected. Finish: multi-layer, zinc phosphate primer with electrophoretic paint finish.

Standard Equipment
Reclining front seats with fine adjustment and adjustable head restraints. Inertia reel front seatbelts, sensitive to both vehicle and passenger movement. Rear seat belt anchorage points. Glove box. Floor console. Cigar lighter. Twin courtesy lights. Sports steering wheel. Fabric upholstery. Felt carpeting with rubber heel mats. Fold flat rear seat. Counter balanced tailgate with gas-filled struts. Removable luggage compartment cover/parcel shelf beneath tailgate. Dipping interior mirror. Vanity mirror. Driver's door mirror. Laminated windscreen. Intermittent wiper control. Electric screenwash. Automatic reversing lamp. Heated rear window. Side repeater flashers. Load adjustable headlamps. Headlamp flasher. Rear fog light. Hazard warning flashers. Low fuel warning light. Low brake fluid warning light. Handbrake warning light. Low oil pressure warning light. Temperature guage. Front and rear towing eyes. Rubber bumper end caps. Multi-piece tool kit. First aid kit. Locking fuel cap.

Fuel Capacity
8.35 gallons.

Fuel Consumption
Simulated urban driving 28.5 m.p.g.
Constant speed driving 56 m.p.h. — 37.6 m.p.g.
Constant speed driving 75 m.p.h. — 28.5 m.p.g.

Kerb Weight
1840 lbs

Turning Circle
33'9" 10.28 metres

Dimensions
Length 149.25 in. 3792 mm.
Width 62.50 in. 1588 mm.
Height 54 in. 1372 mm.

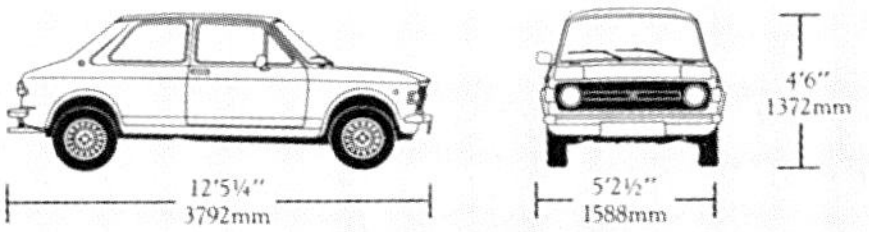

ZASTAVA
CARS

Zastava Cars (GB) Ltd. Worcester House, Basingstoke Road, Reading RG2 0QW. Tel: Reading (0734) 866921.

YOUR LOCAL DEALER
DOUBLE 'S' MOTORS LTD.
KIRBY GARAGE,
CALDY ROAD,
WEST KIRBY.
TEL. 051-625-6186
VAT No. 165 1943 49

The new broom started with the Zastava 1100. This was a hatchback version of the Fiat 128, again never produced in Italy. The 1100 was available in two-door and four-door form, and the booted version was also available – simply a Fiat 128 with minor differences. Exports were quite successful. The Zastava was known in Germany as the Fiat 1100z, while confusingly in Poland it was known as the Zastava 1100p. Curiously the booted version came last, with production only starting in 1984. 1980 saw the company name changed to Yugo, as it was felt that this was more easily pronounceable overseas, and the first home-grown model appeared. This was a Fiat 127 re-skinned and known as the Yugo 45. Power was provided by the Fiat 900cc engine, although 1100cc and 1298cc versions were available, known as the Yugo 55/65.

Styling was quite attractive, although the top-of-the-range model for the British market attempted the boy racer look with white air dams and wide stripes, which gave the machine the appearance of having been tarted up on a Saturday afternoon. The Yugo 65 is quite an attractive drive, albeit somewhat noisy. The fifth gear helps with motorway cruising, while the handling is pure Fiat, giving the impression of how the Mini could have evolved. 1987 saw the launch of the Cabrio.

Also in 1987 the new type 101, known as the Florida, was born. Italdesign produced the body, with engine options being 1400cc, 1600cc and the 1700cc diesel. The designers produced quite a streamlined automobile with 0.32 wind resistance, although the styling could not be described as exciting. The Yugo Florida was imported into the UK as the Sana, but the Bosnian conflict cut short this arrangement fairly quickly – and the Sana remains extremely rare (and cheap). The late '80s were good for Yugo with production reaching 239,000 in 1988. The factory also produced quite a wide range of weaponry for the Serbian army, and the history of Yugo grew to a close in 1999 with heavy allied bombardment closing the factory until early 2000. Amazingly, the most devoted Yugo fans are the Americans, and the Yugo newsgroup on the web is full of banter about the latest hot cam and performance mods. March 2000 also saw the launch of a surreal film, *Drowning Mona*, where an entire town drives Yugoslavia's best.

9

Bulgaria

Bulgaria has never been known for automotive excellence but has arguably produced the Eastern Bloc's best sports car – so please do not turn the page but read on. Those who have thought up names for Eastern Bloc cars have never been known for their originality, and in Bulgaria the Communist authorities remained true to form by announcing that a prototype car produced in Lovec in 1961 was to be known as the Balkan (above). This two-door saloon only ever existed in prototype form, with an engine sourced from the USSR.

The Lovec factory was used to produce Moskviches following a 1966 treaty with the USSR. In 1967 the first Moskvich 408, known as the Rila 1400, rolled off the production line. The Rila range of mountains is well known in Bulgaria. Production amounted to around 15,000 units per year, and soon the 408 model was replaced by the 412 and then the 2410 Moskvich, with slightly more modern lines. Diesel Rilas were produced using a Perkins engine built under licence. While production of the Aleko was planned for 1988 nothing surfaced, and the factory is now defunct.

Fiats were made in Bulgari between 1966 and 1968. Econo mies of scale were clearly not c prime concern with total produc tion never exceeding four figures The Fiat 850, 124 and 124 coup were produced by the Pirin Fia factory. Were there such things a Bulgarian yuppies? If they eve existed, they must have bee attracted to Balkan Renault' Alpine coupé and Alpine cabriole (left) produced between 1967 an 1971. Other Renaults in produc tion included the 8 and 10 (th Rally Balkan Renault 10 is show below). Total production ma have amounted to around 3,50(Without a trace of irony, sta literature proudly announced the seats of the Alpine coupé a cabriolet were Bulgarian made The Bulgarian motor industry spluttered back into life in the early 1990s with production of CKD Maestros. Th Rodacar factory in Varna still exists but production has stopped because of heavy government taxes on importe goods. The Maestro is to full British specification with the exception of the suspension, which comes from th Maestro diesel van – in other words, beefed up for Bulgarian roads. In 1999 an enterprising West Midlanc businessman started reconstituting Maestros from CKD kits that had never been delivered, and a number wer sold for £4,000 in Spain. Britain's cheapest car in 1999 was the rhd version, sold for just over £5,000 by Parkwa Service Station, Ledbury.